Principles of Air Pollution Meteorology

Principles of Air Pollution Meteorology

T.J. Lyons and W.D. Scott

Belhaven Press
A division of Pinter Publishers
London

First published in Great Britain in 1990 by
Belhaven Press (a division of Pinter Publishers),
25 Floral Street, London WC2E 9DS

Note: Figures 17.1 and 17.6 from *Fundamentals of Air Pollution* by Arthur Stern, Henry Wohlers, Richard Boubel, and William P. Lowery, copyright © 1973 by Harcourt Brace Jovanovich, Inc., reproduced by permission of the publisher.

British Library Cataloguing in Publication Data

A CIP catalogue record for this book is available from the
British Library

ISBN 1-85293-079-9

Filmset by Mayhew Typesetting, Bristol
Printed in Great Britain by Antony Rowe Ltd, Chippenham, Wiltshire

CONTENTS

ACKNOWLEDGEMENTS

Credit for the material contained herein should go to countless individuals who, over time, have contributed to the science of air pollution meteorology. Many figures in the book have come from other sources and are presented here with the permission of the American Meteorological Society, the American Society of Testing Materials, Academic Press, Kluwer Academic Publishers, the US Department of Energy, the Electric Power Research Institute, the University of Chicago Press, D. Reidel Publishing Company, Elsevier Science Publishers, Pergamon Press and Springer-Verlag.

Chapter 1

INTRODUCTION

The atmosphere contains a mixture of gases and solid particles ranging in size from less than one thousandth of a millimetre up to a millimetre or more. This mixture is affected by continuous, dynamic change with vast quantities being added to or removed from the atmosphere by various natural and industrial processes. Our focus is on industrial processes but pollutants are continuously being raised or moved by the wind and volcanoes, transformed by radiation or reactions with water, or cycled variously through our biological environment.

Many of these constituents are vital to our existence, some are harmful and others are benign. Pollutants are normally defined as those substances which are poisonous to humans, animals or vegetation, have an objectionable odour or irritate our senses, obscure visibility or damage property. Such pollutants (see Table 1.1) may result from natural processes or be the direct result of our activities.

Air pollution meteorology is concerned with the fate of these pollutants once they are emitted into the atmosphere and addresses natural releases, deliberate industrial emissions and accidental spills. This book usually presents examples of industrial emission but the techniques developed are equally applicable to natural sources. In this context, it is important to note that we are not concerned with engineering solutions at the source but rather in developing ways of predicting the resulting atmospheric consequences.

The ultimate solution to industrial air pollution problems lies at the source. Still, there will be pollution, despite the best engineering solutions. Even if all industries were brought to the zero emission level, there is the possibility of accidental spills. In our daily life we pollute when we eat, breathe and drive motor cars. Nature continuously adds terpenes, smoke,

Table 1.1 Examples of air contaminants

Group	Specific examples
Solids, fine (< 76 μm) and coarse (> 76 μm)	Carbon, fly ash, zinc oxide, silicates, resins, sulphates, fluorides, metallic dust, tars, chlorides.
Sulphur compounds	Sulphur dioxide, sulphur trioxide, hydrogen sulphide, mercaptans.
Organic compounds	Aldehydes, hydrocarbons, tars.
Nitrogen compounds	Nitric oxide, nitrogen dioxide, ammonia.
Oxygen compounds	Carbon monoxide, carbon dioxide, ozone.
Halogen compounds	Hydrogen fluoride, hydrogen chloride.
Radioactive compounds	Radioactive gases, aerosols, etc.

Source: After Gilpin, 1971.

volcanic materials and meteorites. Also, ultimate engineering solutions may be of questionable economic viability, and a clear understanding of the inherent environmental risk needs to be appreciated if these solutions are not adopted.

Pollutants emitted into the atmosphere undergo transport, dispersion, transformation and ultimate removal from the atmosphere (Seinfield, 1986). Transport is through the action of the mean wind in carrying the pollutant away from its source, whereas dispersion results from the turbulent characteristics of the atmosphere in diffusing the pollutant in all directions. Since most industrial pollutants are injected into the atmosphere near the surface of the earth, the physics of this atmospheric boundary layer controls the transport and dispersion of pollutant.

Transformation changes the nature of a pollutant species. This may involve a chemical reaction between two atmospheric pollutants, a reaction of the pollutant with some common constituent of the atmosphere (e.g. CO_2, H_2O, N_2), or reaction involving some other factor, such as sunlight in the case of the formation of photochemical smog. Our understanding of atmospheric chemistry is far from complete. Still, here we are mainly interested in identifying the chemicals and other pollutants and how they move about. Ultimately, with an understanding of the whole we can generate an air-quality model.

Of course, pollutants are ultimately removed from the atmosphere through gravitational settling, interaction with surface features such as buildings, plants or topographical features, or washout through precipitation. Once again our understanding is wanting but the fate of the pollution cannot be ignored in our final model development.

In surveying air pollution we consider primarily the physical processes in the atmospheric boundary layer. These processes lead to an understanding of the transport and diffusion of atmospheric pollutants and a general appreciation of the concepts of advection and turbulence. This is the start of the pollutant pathway; inclusion of the chemical nature of the pollutants and incorporating of known transformation and removal processes completes the pollutant pathway. We finally put together an air-quality model which allows the full tracing of the pathway if only in terms of average properties.

The accuracy and usefulness of an air-quality model will depend on the inherent limitations of the model and the accuracy of the input data. Increasing the accuracy of the input data will not necessarily increase the accuracy of the model unless the model is particularly sensitive to that data, but the input data must be representative of the region under study and appropriate to the model. Thus we are led to a consideration of the requirements of environmental monitoring. This monitoring encompasses both physical measurements and pollutant impact and introduces the concept of exposure-damage. Damage can even be used as a measurement tool. With the air-quality model, we are linked to some understanding of environmental risk.

Of necessity we will concentrate on the development of air-quality models for lighter than air gases and the application of these to outdoor environmental impacts. This is purely an arbitrary division but tends to neglect developments in understanding the dispersion of denser-than-air gases (such as Blackmore *et al.*, 1982; Ermak *et al.*, 1982; Woodward *et al.*, 1982; Koopman *et al.*, 1989) and the impact of indoor air pollution (Wadden and Scheff, 1983; Sexton and Wesolowski, 1985), although the techniques developed are applicable to these areas.

Air-quality models form the first step towards an understanding of the impact of air pollutants (see Figure 1.1) and we hope they will remove some of our ignorance of the pathway of pollutants through the atmosphere. We do not consider exposure-damage and value models which attempt to assess the full environmental impact of air pollutants with particular respect to health (see Table 1.2) and environmental risk assessment issues. Nevertheless, the concepts and issues are introduced.

Thus we are interested in laying the foundations for an understanding of the pollutant pathway processes through the atmosphere, keeping in mind our overall aim to develop techniques capable of assessing the full environmental impact of any emission in terms of its environmental cost. An evaluation of the total environmental cost enables an appropriate assessment of the associated environmental risk and leads to a value judgement as to the acceptability or otherwise of that risk. Thus we seek to lay the foundations of air-quality modelling as an essential component of environmental science through a discussion of the pathway processes in the atmosphere and the estimations of ambient levels.

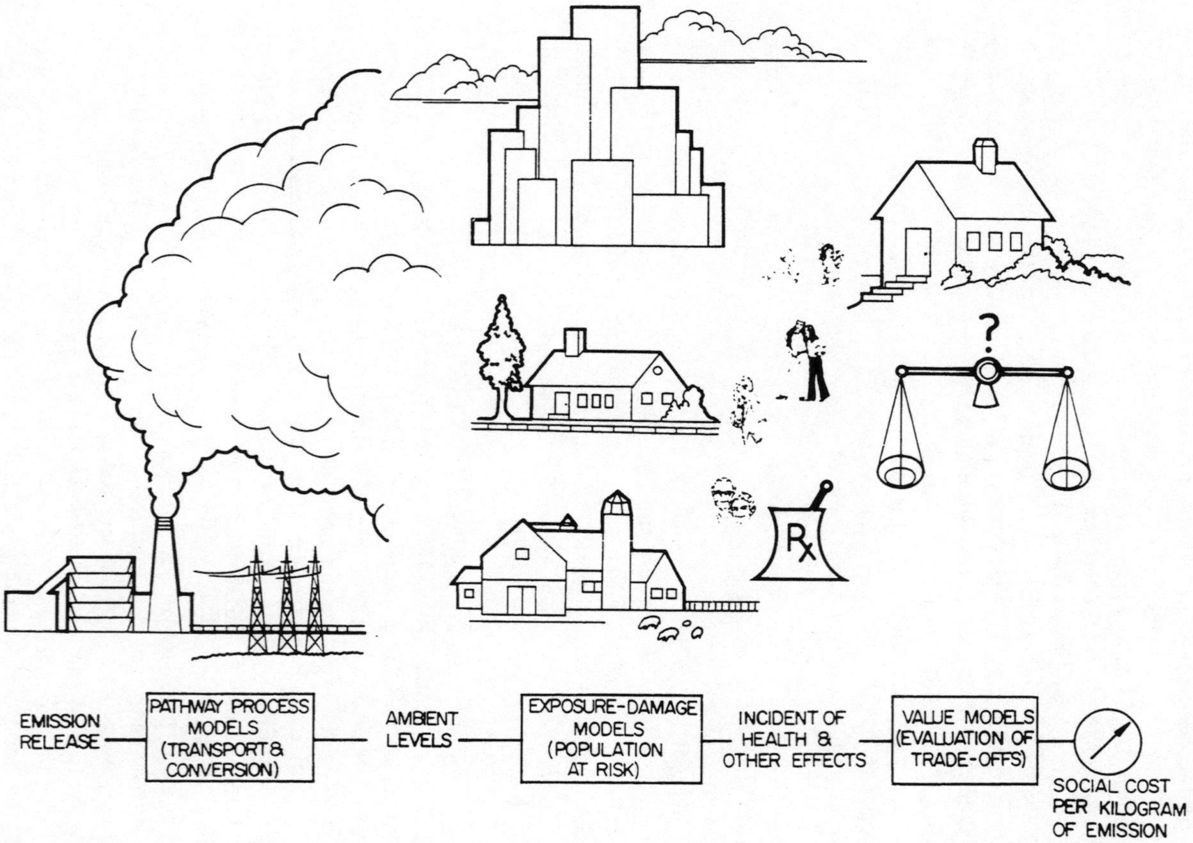

Figure 1.1 Causal chain of models from emission to consequences of airborne pollutants (after Drake *et al.*, 1979) Copyright © 1979 Electric Power Research Institute. Reprinted with permission.

Table 1.2 Air pollutants with recognised or potential long-term effects on health at usual air pollution levels

Substances with known effects on health (acute or chronic)	Substances thought to have long-term effects *per se**	Potential long-term effects of combinations
Arsenic	Arsenic (arsenical dermatitis)	
Asbestos	Asbestos (asbestosis, mesothelioma)	
Beryllium	Beryllium (berylliosis)	Be + F (fluorides potentiate pulmonary changes in berylliosis)
Carbon monoxide		Synergistic in pO_2 depression
Carcinogens		Carcinogens produce tumours in presence of promoters
Fluoride	Fluoride (fluorosis)	Fluoride (promotes or accelerates lung disease)
Hydrocarbons		HC + O_3 → tumorigen + influenza → cancer
Hydrogen sulphide (possibly with mercaptans)		Antagonizes pollutants (strictly speaking not detrimental to health)
Inorganic particulates	Inorganic particulates (pulmonary sclerosis)	
Lead		
Nitric oxide		
Nitrogen dioxide	Nitrogen dioxide (mild accelerator of lung tumours)	NO_2 + micro-organisms (pneumonia) + HNO_3 (bronchiolitis, fibrosa obliterans) + tars (smoker's lung cancer)
Organic oxidants (peroxyacylnitrates)		
Organic particulates (asthmagenic agents)	Asthmagenic agents (asthma)	
Ozone	Ozone (chronic lung changes, accelerated aging)	O_3 + micro-organisms (lung-tumour accelerator)
Sulphur dioxide, Sulphur trioxide		SO_2, SO_3 + particulates aggravate lung disease

*Effects are given in parentheses

Source: After Gilpin, 1971.

If the environment was a static entity, such estimates would be sufficient to define future trends. However, the natural environment is undergoing a continual evolution and it is important to identify existing trends so that these can be accounted for in predicting likely future air quality. Such trends are addressed through the impact of pollutants on the environment in defining air-quality indices that can act as a guide to both natural and induced changes.

Although our understanding of the pathway processes in the atmosphere and our ability to model them is far from complete, we cannot await full knowledge before attempting to estimate the environmental consequences of proposed activities. Thus we are interested in establishing the best practical means of assessing pollutant concentrations bearing in mind the limitations of our understanding. The development of this understanding is described extensively elsewhere (i.e. Slade, 1968; Pasquill, 1974; Haugen, 1975; Nieuwstadt and van Dop, 1982; Randerson, 1984; Venkatram and Wyngaard, 1988) and we shall draw on that material in discussing the basic principles.

Chapter 2

THE ATMOSPHERIC BOUNDARY LAYER

The fate of atmospheric pollutants is affected by atmospheric flows or advection. In turn, the flows are brought about by movement of energy through the atmosphere and the interaction of the atmosphere with the underlying surface. The atmospheric boundary layer is the layer of the atmosphere that comes under the direct influence of the earth's surface. The boundary layer is where we live. It contains the air we breathe and it is where we pour our pollutants. Air motions within it are retarded by the frictional drag of the earth's surface and the heat and moisture structure are determined to a large extent by energy exchanges that take place at the surface.

2.1 Solar radiation

Solar radiation is received by the earth's surface and transferred to the atmosphere in the form of sensible heat, latent heat and long-wave radiation. The sun is the direct energy source but this transfer of heat from the solid earth and the oceans is the second link of a chain process that drives the local and larger-scale weather systems. On the local scale, the sea breeze is a direct consequence of the uplift resulting from preferred heating of the land and subsequent heating of the overlying air. On the larger scale, heat associated with water vapour, latent heat, is transferred to the air from the warm, moist ocean surface of the tropics. This latent heat drives a few thousand gigantic cumulonimbus clouds in the intertropical convergence zone, and these clouds are considered the engine that drives the global circulation.

Hence, we consider the mechanism of heat input before proceeding with the advective and diffusive transport of atmospheric pollutants. Radiation from the sun is in the form of electromagnetic radiation, which is energy derived from oscillating magnetic and electrostatic fields. It is one form of energy capable of being transmitted through space. This radiation from the sun is, of course, dependent upon the time of day, the season, and masking influences such as cloudiness. It also varies with wavelength; this dependence may be approximated by Planck's distribution law of emission:

$$E_\nu \mathrm{d}\nu = \mathrm{C}_1 \nu^3 (\exp(\mathrm{C}_2 \nu/T) - 1)^{-1} \mathrm{d}\nu \qquad (2.1)$$

where $E_\nu \mathrm{d}\nu$ is the amount of radiant flux (energy area $^{-1}$ time^{-1}) emitted by a black body at absolute temperature T in the frequency band ν to $\nu + \mathrm{d}\nu$. The constants in this equation are

$$\mathrm{C}_1 = \frac{2\pi h}{c^2} \text{ and}$$

$$\mathrm{C}_2 = \frac{h}{k},$$

where, in the SI (Système International) system,

$h = 6.63 \times 10^{-34}$ Joule s (Planck's constant),
$c = 3.0 \times 10^{8}$ ms^{-1} (speed of light), and
$k = 1.37 \times 10^{-23}$ Joule deg^{-1} (Boltzmann's constant).

Planck's distribution law of emission is defined for a black body, which is a material whose properties are describable in theory but which, as far as is known, does not exist. A black body absorbs all the electromagnetic radiation that is incident upon it and is best described physically by a hole in a cavity. The sun may be approximated by a black body at a temperature of 6,000° K.

The amount of radiation absorbed by a body defines the absorptivity, a, of that body, which is simply the fraction of incident energy that is absorbed. Consequently, the absorptivity of a black body is unity, the maximum possible. It is also possible to define an emissivity factor, ϵ, as the ratio of energy emitted by an object or material surface to the energy emitted by a black body at the same temperature and frequency. Hence by definition the emissivity of a black body is unity; all other bodies are generally termed grey bodies and have emissivities less than unity. A corollary is that a black body emits the maximum amount of energy physically possible for any combination of frequency and temperature.

Kirchhoff's law states that the absorptivity of a surface is equal to the emissivity for the same frequency and temperature. That is, a good

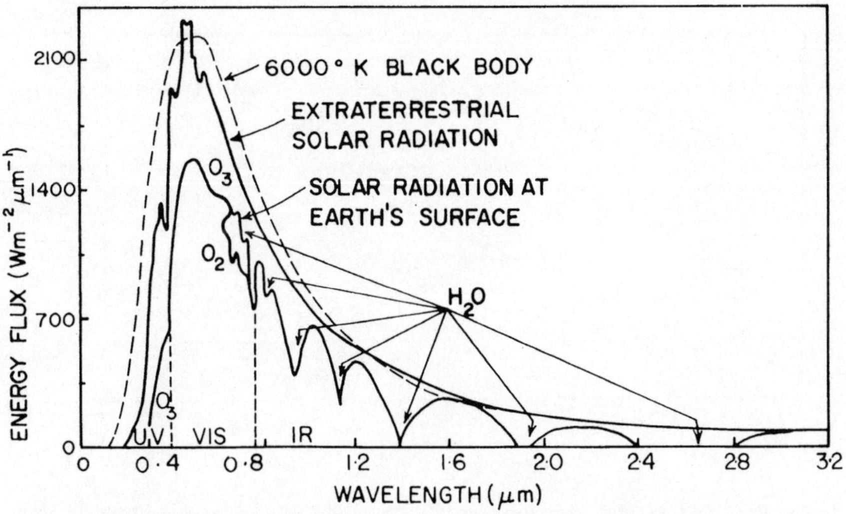

Figure 2.1 Emission spectra of the sun (after Kondratyev, 1969)

absorber is also a good emitter and

$$[a_\nu = \epsilon_\nu]_T \tag{2.2}$$

Hence emission spectra are also absorption spectra; the emission spectra of the sun are shown in Figure 2.1. Comparison of this figure with the theoretical emission spectra of a black body, obtained from Planck's distribution law, illustrates that the sun is not a perfect black body.

The frequency ν and wavelength λ of radiation are related by the expression

$$\nu\lambda = C \tag{2.3}$$

and, according to Planck, the energy of a photon and thus the energy of a single photon or packet of light is proportional to the frequency and given by,

$$E = h\nu \tag{2.4}$$

Hence the energy of a photon is directly proportional to frequency and inversely proportional to wavelength. Of course, the energy flux depends not only on the energy per photon but also the number of photons passing per second.

Plotting Planck's distribution on a linear scale of frequency allows that a unit area on the diagram represents the same amount of energy regardless

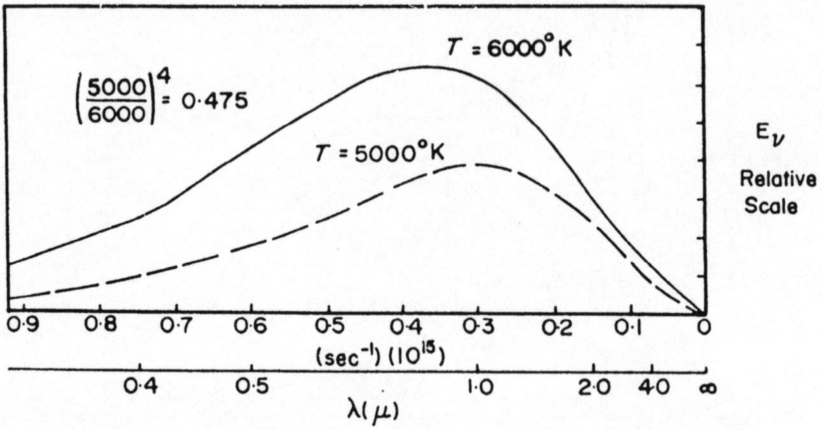

Figure 2.2 Black-body emission spectra for bodies of temperature 6,000°K and 5,000°K (based on Stern *et al.* 1973)

of position. The black-body emission spectra for bodies of temperature 6,000°K and 5,000°K are shown in Figure 2.2. This figure and Planck's analytical expression give a direct measure of the fraction of radiant energy associated with a given wavelength. As can be seen from the figure, the frequency of maximum emission is a direct function of the temperature. This tendency was found empirically by Wien and is called Wien's displacement law; it can be validated by differentiating equation 2.1 to yield Wien's displacement law:

$$\frac{\nu_{max}}{T} = 5.9 \times 10^{10} \text{ s}^{-1} \text{ deg}^{-1} \qquad (2.5)$$

or, in terms of wavelength,

$$T\,\lambda_{max} = 2{,}897\ \mu \text{ deg.} \qquad (2.6)$$

This result illustrates the displacement of the wavelength of maximum emission with the temperature of the emitting surface. That is, light of short wavelength radiation is received by the earth from the sun, but the earth, emitting at a much cooler temperature (of the order of 300°K) emits long wavelength radiation. At the same time the atmosphere is relatively transparent to short-wave radiation and somewhat opaque to long-wave radiation. This results in a trapping effect that raises the earth's temperature appreciably and is known commonly as the greenhouse effect.

The energy reaching the earth is affected by the spectral emissions from the sun and the absorption characteristics of the atmosphere (see Figure 2.1). The spectral envelope shows the influence of each individual constituent with many peaks and 'windows'. Still, as a first approximation, we follow the simple model of the sun as a black body with an effective temperature of 6,000°K (see the dashed line on Figure 2.1). The total radiant flux (irradiance) emitted by the sun is found by summing over all wavelengths, or integrating equation 2.1:

$$F = \int_0^\infty E_\nu \mathrm{d}\nu$$

$$= \int_0^\infty \mathrm{C}_1\nu^3 \left(\exp\left(\frac{\mathrm{C}_2\nu}{T}\right) - 1\right)^{-1} \mathrm{d}\nu$$

Substituting for the constants and defining

$$x = \frac{h\nu}{kT},$$

this equation may be written as

$$F = \frac{2\pi h}{c^2}\,\frac{k^3T^3}{h^3}\,\frac{kT}{h}\int_0^\infty \frac{x^3}{\mathrm{e}^x - 1}\,\mathrm{d}x$$

and the integral has the value $\pi^4/15$,

$$\text{i.e. } F = \int_0^\infty E_\nu \mathrm{d}\nu = \frac{2\pi^5k^4T^4}{15c^2h^3}$$

$$\text{or } F = \sigma T^4, \tag{2.7}$$

which is the Stefan–Boltzmann law; the Stefan–Boltzmann constant, σ, has the value 5.67×10^{-8} W m^{-2} K^{-4}. As the sun can be approximated by a black body at a radiating temperature of 6,000°K, the total solar flux is approximately

$$(4\pi R_s^2)\,\sigma\,T_s^4 = 3.9 \times 10^{26} \text{ W},$$

where R_s is the sun's radius, and T_s is the effective temperature of the sun's surface. By the inverse square relationship of diminishing intensity (all the sun's radiation passes outwards through ever more increasing spherical shells of radius $4\pi R^2$) this flux is reduced to 1,360 Wm^{-2} after travelling the 1.5×10^8 km to reach to outer limits of the earth's atmosphere. This radiant flux is known as the solar constant and is thought to be nearly invariant with time. Since the spherical area of the earth's atmosphere is four times the area exposed to the perpendicular rays of the sun, the average solar flux over the entire surface of the earth is 340 Wm^{-2}.

The earth is not a black body, however, and only absorbs a portion of this incoming radiation. On average the earth absorbs 66 per cent of the incoming radiation (i.e. 224 Wm^{-2}) and reflects 34 per cent. In meteorological terms it is said that the earth has a reflectivity or global albedo, as measured at the top of the atmosphere of 0.34. It must be emphasised that this is an average global value and the albedo of any surface is dependent on the surface characteristics. For example, the albedo of freshly fallen snow is approximately 0.85 whereas that for dark fir tree forests is less than 0.10. The effective radiant temperature of the earth–atmosphere system can be estimated using the Stefan–Boltzmann law and Kirchhoff's law and the resulting temperature is

$$\sqrt[4]{\frac{(340)\ (0.66)}{\sigma}} = 252°\text{K}.$$

Solar radiation incident on an atmospheric layer can be transmitted, reflected or absorbed. Hence is it normal to define an absorptivity (a) as well as a reflectivity (r) and a transmissivity (τ) such that

$$a + r + \tau = 1.0. \tag{2.8}$$

Each of these quantities depends upon the wavelength and may be defined for a single wavelength, a waveband or as an average over the entire spectra.

If a layer of the atmosphere of thickness dz is considered the transfer medium, it has been found experimentally that the fractional decrease in flux intensity is proportional to the optical path taken by the light (see Figure 2.3). This is a net result of absorption of the radiation and multiple scattering by aerosol particles, droplets and gases and can be represented by

$$\frac{dI_\lambda}{I_\lambda} = -\ k_\lambda \sec \theta \ dz, \tag{2.9}$$

where I_λ is the incident radiant flux or irradiance and θ is the angle of the beam from the normal to the atmospheric layer or the zenith angle. The quantity (sec θ dz) is the optical path and k_λ is the absorption or extinction coefficients (units of length^{-1}) for the medium at the wavelength in question. This coefficient is the sum of the air's absorption and scattering coefficients (Butcher and Charlson, 1972) and is proportional to the concentration of scatterers or absorbers present in the air.

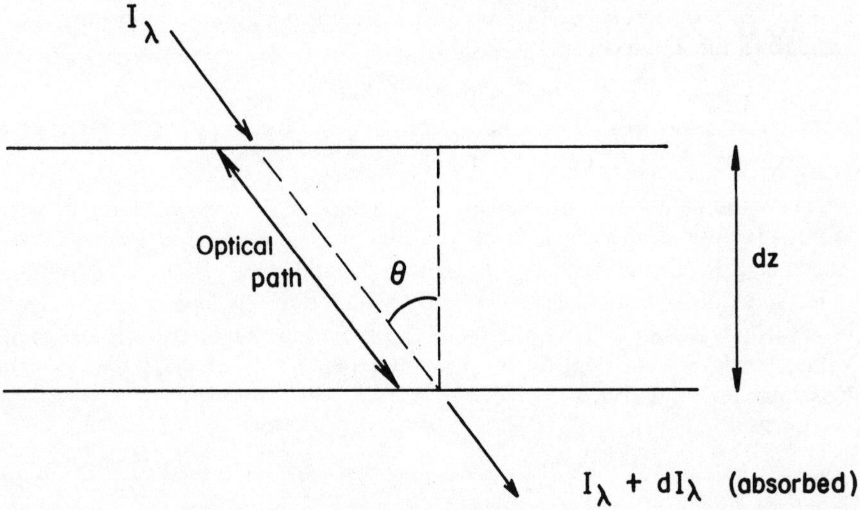

Figure 2.3 Attenuation of light through an atmospheric layer

If k_λ is considered to be independent of z, the differential form can be integrated through the whole depth of the atmosphere Z. Then the fraction of the incoming flux that reaches the ground is given by

$$\frac{I}{I_0} = \exp\,(-k_\lambda\ Z \sec\,\theta),$$

which is the normal Beer's law form for the case when light is transmitted obliquely. Here the quantity $\sec\,\theta = M$ is called the 'air mass'; the remaining exponential terms are called the 'optical thickness', T (McCartney, 1976);

$$T_\lambda = k_\lambda Z.$$

This is for the case when the properties of the atmosphere are assumed to be uniform, which is never the case in practice. Going back to the differential equation, we see that we could just as well have allowed k_λ to vary. Then integration is over the whole depth of the atmosphere,

$$\text{i.e.} \quad T_\lambda = \int_{\text{surface}}^{\text{top of atmosphere}} k_\lambda\ dz$$

However, when only single scattering is concerned, it is the total number of molecules or aerosols, and not their distribution, that is important. Then we define the reduced height Z',

$$Z' = \frac{P_0}{\varrho_0 g} = \frac{RT_0}{g},$$

which is the height the atmosphere would have if it were constant density. With this approximation, it is possible to use the integrated form directly, knowing a mean value of k_λ through the whole depth of the atmosphere.

If the extinction coefficient can be considered as the inverse of the depth of atmosphere required to attenuate the incoming light level to 1/e of its value; then it is related to the meteorological range, L_ν, by the Koschmieder equation,

$$L_\nu = \frac{3.9}{k_\lambda}, \qquad (2.10)$$

which is very useful as a rough estimate of pollution (Butcher and Charlson, 1972; Richard, 1988).

2.2 Terrestrial radiation

As we have seen, the average global absorption of the earth–atmosphere system is 224 Wm^{-2}. Since absorptivity and emissivity are equal, the earth–atmosphere has an average global emission of 224 Wm^{-2}. From the Stefan–Boltzmann law, this is emitted at an effective radiant temperature of 252°K. This temperature is much lower than the effective radiant temperature of the sun and the Wien displacement law implies that terrestrial radiation has its wavelength of maximum emission at a longer wavelength than solar radiation. Hence terrestrial radiation is often referred to as long-wave radiation.

The long-wave radiation received at the ground is a net result of radiation from the sky and radiant energy emitted by the earth's surface. The atmosphere, since it is not at absolute zero temperature, emits radiant energy in the long-wave region and exhibits an effective emissivity. This flux of long-wave radiation from the lower atmosphere may be obtained from appropriate radiometers or from empirical equations developed from such data. The long-wave flux from a cloudless sky has been related by various authors to the heat and moisture content of the lower atmosphere.

For example, Angstrom (1916) found that

$$I_i = \epsilon \, \sigma \, T^4 \, (a_0 - b_0 \exp(-2.3 \, c_0 e)), \qquad (2.11)$$

Table 2.1 Empirical factor to account for cloud type in estimating incoming long-wave radiation

Cloud	Altitude (typical) metres	K
Cirrus	12,200	0.04
Cirrostratus	8,400	0.08
Altocumulus	3,700	0.16
Altostratus	2,100	0.20
Stratocumulus	1,200	0.22
Stratus	500	0.24
Nimbostratus	100	0.25

Source: Geiger, 1965.

where I_i is the counter radiation (incoming long-wave radiation) from clear skies; a_0, b_0, c_0 are empirical constants, T is the air temperature near the surface in K, and e is the vapour pressure of the air near the surface in hPa. Listed values of a_0 range from 0.710 to 0.820, b_0 from 0.148 to 0.326 and c_0 from 0.041 to 0.094 (Sellers, 1965).

Brunt (1932), on the other hand, found that the simple relationship

$$I_i = \sigma T^4 (b + c e^{1/2}) \qquad (2.12)$$

fitted his observations better than the Angstrom equation. His empirical constants b and c have listed values ranging from 0.34 to 0.71 for b and 0.023 to 0.110 for c (Sellers, 1965).

The presence of clouds may be accounted for empirically by adding a factor for cloudiness (Geiger, 1965). In this instance,

$$I_{i,\mathrm{cloud}} = I_i (1 + Kn^2), \qquad (2.13)$$

where K is related to the height and type of cloud and typical values are given in Table 2.1; and n is the amount of cloud cover in tenths.

Radiation emitted by the earth's surface is given by

$$I_0 = \epsilon \sigma T^4,$$

then the net long-wave flux at the earth–atmosphere interface, using the Brunt formula is

$$I_i - I_0 = \epsilon \sigma T^4 ((b + c e^{1/2}) (1 + K n^2) - 1). \qquad (2.14)$$

In dry air this net flux of outgoing radiation with clear skies rarely exceeds

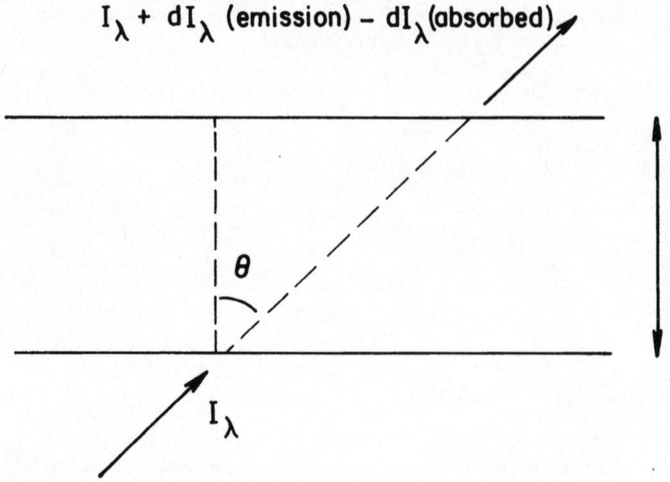

Figure 2.4 Absorption and emission of long-wave radiation

30 per cent of the total infrared radiation sent out by the earth's surface (i.e. I_0) whereas under conditions of large vapour pressure it is less than 20 per cent of I_0. In the absence of an atmosphere this net upward long-wave radiation would be lost directly to space. However the atmosphere absorbs about 70 per cent of I_0 (under clear skies) re-radiating it in all directions. That is, the atmosphere's short-wave transmissivity is greater than its long-wave transmissivity.

Following Beer's law, the absorption of terrestrial radiation along an upward path through the atmosphere can be described as

$$dI_\lambda \text{ (absorption)} = I_\lambda k_\lambda \varrho \sec \theta \, dz = I_\lambda \, da_\lambda \qquad (2.15)$$

where ϱ is the density of the gas.

As noted previously, a layer of the atmosphere also emits long-wave radiation of its own, following the Stefan–Boltzmann law. That is, it acts as a radiating body at temperature T and its emitted radiation is

$$dI_\lambda \text{ (direct emission)} = I_\lambda{}^* \, d\epsilon_\lambda \qquad (2.16)$$

where $I_\lambda{}^*$ (= σT^4) is the black-body monochromatic radiance specified by Planck's law. However, from Kirchhoff's law, absorptivity and emissivity are equivalent and therefore

$$dI_\lambda \text{ (direct emission)} = I_\lambda{}^* \, da_\lambda.$$

Subtracting the absorption from the emission gives the net contribution of the atmospheric layer to the monochromatic radiance of the long-wave radiation passing upward through it. That is, dI_λ (total) = dI_λ (direct emission) − dI_λ (absorbed)

$$
\begin{aligned}
&= I_\lambda^* \, da_\lambda - I_\lambda \, da_\lambda \\
&= (I_\lambda^* - I_\lambda) \, da_\lambda \\
&= k_\lambda \varrho \sec\theta \, dz \, (I_\lambda^* - I_\lambda) \\
dI_\lambda &= -k_\lambda (I_\lambda - I_\lambda^*) \, \varrho \sec\theta \, dz. \qquad (2.17)
\end{aligned}
$$

This expression, known as Schwarzchild's equation, is the basis for the computation of the transfer of infrared radiation. For an isothermal gas with constant k_λ, equation 2.17 may be integrated to obtain

$$I_\lambda - I_\lambda^* = (I_{\lambda 0} - I_\lambda^*) \exp(-\sigma_\lambda) \qquad (2.18)$$

where σ_λ is the optical depth, $k_\lambda \varrho \sec\theta \, z$, and $I_{\lambda 0}$ is the radiance incident on the layer from below. This expression shows that I_λ should exponentially approach I_λ^* as the optical thickness of the layer increases. For a layer of infinite optical thickness the emission from the top is I_λ^* regardless of the value of I_λ; that is, such a body behaves as a black body.

We are now in a position to consider a simplified model of radiative flux in an earth–atmosphere–space system (Stern *et al.*, 1973). Let us assume that the earth is a black body (i.e. $a_e = 1.0$) and that the atmosphere is semitransparent but not reflective (i.e. $\tau_a = 1 - a_a$). The reflective property of the real atmosphere is accounted for by decreasing the incoming solar flux from space. Also the model omits the nonradiative transfer which takes place by convection between the earth and the atmosphere.

	Short-wave	Long-wave	
Space	$S^1 = (1 - \tau_{as}) S$	$\tau_{aL} F_e$	$\tau_{aL} F_a$
Atmosphere	$(1 - \tau_{as}) S^1$	$(1 - \tau_{aL}) F_e$	$2 \tau_{aL} F_a$
Earth	$\tau_{as} S^1$	F_e	$\tau_{aL} F_a$

Source: after Stern *et al.*, 1973

Figure 2.5 Fluxes of radiation in the atmosphere

Figure 2.5 shows the fluxes of short- and long-wave radiation and the amounts absorbed and transmitted by the individual sections of the model. Balancing incoming radiation against outgoing radiation we find that at the space interface

$$\tau_{aL} F_e + \tau_{aL} F_a = S^1$$

within the atmosphere,

$$(1 - \tau_{as}) S^1 + (1 - \tau_{aL}) F_e = 2\tau_{aL} F_a,$$

and at the surface of the earth,

$$\tau_{as} S^1 + \tau_{aL} F_a = F_e.$$

Considering only the radiation balance at the surface of the earth, and eliminating F_a, leads to

$$S^1 (1 + \tau_{as}) = F_e (1 + \tau_{aL})$$

$$\text{or } \frac{F_e}{S^1} = \frac{(1 + \tau_{aS})}{(1 + \tau_{aL})} = \frac{\sigma T_e^4}{S^1}. \tag{2.19}$$

Since by definition, $0 < \tau_{aS}, \tau_{aL} < 1$ and in the real atmosphere $\tau_{as} > \tau_{aL}$, the flux density from the earth's surface exceeds that entering the atmosphere. The more τ_{as} exceeds τ_{aL}, the greater will be F_e relative to S^1 and thus the larger will be the mean temperature of the earth's surface, resulting in the greenhouse effect. In other words the 'blanket' effect of the earth's atmosphere that traps heat energy results from the atmosphere being more transparent to short-wave radiation (higher τ_{as}) and less transparent to long-wave radiation (lower τ_{aL}).

One of the more important features of radiant heat transfer in the atmosphere, at least as far as air pollution studies are concerned, is the radiation inversion. This formation can be simulated by considering a series of overlying layers, starting at the surface with the surface layer losing 1 per cent of its starting heat to space, 9 per cent to the layer above, and all other atmospheric layers losing 9 per cent to the layer above and 9 per cent to the layer below. In this simulation only the surface layer loses heat to space and the surface layer neither loses heat nor gains heat from the underlying soil. This indirectly simulates the loss of heat from the lower layer due to contact with the cooling ground (Stern *et al.*, 1973).

Table 2.2 shows the amount of heat in each layer and illustrates the formation of a radiation inversion, which usually occurs in relatively stagnant air at night when the chief mode of heat transfer is by long-wave radiation at and near the earth–atmosphere interface. Through the interworkings of the Stefan–Boltzmann and Kirchhoff laws, the differential absorptivities of air and solid materials at the interface, the air nearest the interface cools first and most rapidly, followed by the layer of air just above, and then by the layer of air next above that, and so on as long as

Table 2.2 Simulation of nocturnal radiation inversion formation

		Time step (approximate hours)		
Layer	Initial	1	3	5
4	1,000.0	1,000.0	1,000.0	1,000.0
3	1,000.0	1,000.0	1,000.0	1,000.0
2	1,000.0	1,000.0	1,000.0	999.4
1	1,000.0	1,000.0	997.6	993.0
Surface	1,000.0	990.0	972.8	958.4

Source: after Stern *et al.*, 1973

skies are clear, winds are calm and until the sun rises and begins to heat the ground.

2.3 Soil temperature

The earth's surface reacts in a special way to the exchange of heat; it does not simply remain at a constant temperature. The cooling of the surface results in an upward flow of heat from the ground, down the temperature gradient. The largest temperature gradients occur within ± 1 cm of the surface and may exceed 5°C cm^{-1}. Even over the ocean a temperature change of 0.5°C in the top millimetre of water may be realised because of evaporative cooling.

The soil is not such a perfect conductor that only small temperature changes can persist; nor is it as insulating towards heat exchange as the air. Still, there is a tendency for the trends shown in Figure 2.6 to appear in daytime (insulation) and night time (nocturnal) situations. Typical temperatures during a summer day are given in Table 2.3. The earth's surface receives or loses most of the radiation and the adjoining atmosphere and soil tend to heat or cool, following the ground temperature. Meteorological screens are normally placed a metre above ground level so that screen air temperature is much lower than the apparent ground temperature during the daytime. At night, screen temperatures are several degrees higher than apparent ground temperatures. The effect varies with type of surface as is shown in Table 2.3, with bare soil becoming much warmer than a grass or water surface on a summer day. Under extreme conditions, phenomenal temperatures occur at the ground surface. For example, surface temperatures in excess of 65°C (338°K) have been observed in outback Western Australia.

The soil temperature results from the overall balance of heat at the earth's surface,

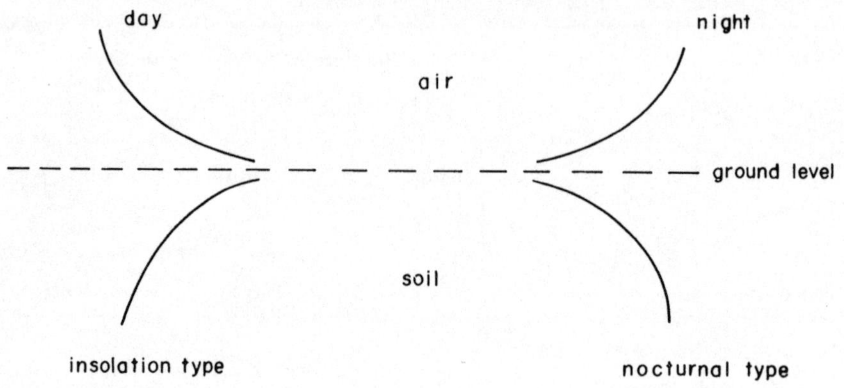

Figure 2.6 Typical temperature profile expected from surface heating and cooling

Table 2.3 Typical temperatures for a summer day, Perth, Western Australia

	Maximum °C	Minimum °C
Screen air temperature (1 metre)	39.9	25.1
Apparent surface temperature		
Short grass	32.0	21.4
Evaporation tank	34.1	19.1
Bare soil	47.2	18.0

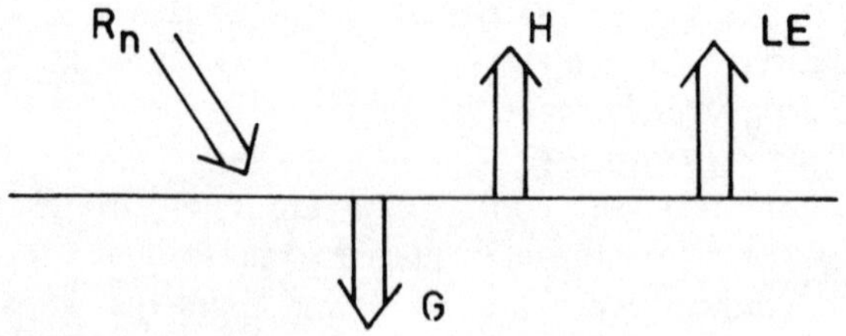

Figure 2.7 Heat balance at the surface of the earth

$$R_n = H + LE + G \tag{2.20}$$

where R_n is the net all-wave radiation received by the surface, H, the sensible heat flux, is the heat directly resulting in temperature change to the atmosphere, G is the ground heat flux, the heat directly resulting in temperature change to the ground, and LE, the latent heat flux, is the

Table 2.4 Layers within the lowest part of the atmosphere

Layer	Vertical extent	Type of transport
Molecular boundary layer	Less than 1 mm adjacent to surface	Molecular conduction and diffusion
Surface layer	Lowest few tens of metres	Microscale turbulence
Mixed layer	About 1 km	Thermals, rolls

Source: after Munn, 1966.

equivalent heat exchange that occurs from the evaporation or condensation of water vapour at the surface.

The evaporation of water at the surface takes the latent heat of condensation with it. The fluxes of heat shown in Figure 2.7 are brought about by temperature and moisture gradients and convective motions in the atmosphere. In this regard the lower portion of the atmosphere is divided according to the type of transport that is dominant. Table 2.4 summarises this layered structure ascribed to the atmosphere.

Within the molecular boundary layer fluid motions are strongly suppressed so that molecular conduction and diffusion are the principal mechanisms for vertical transport of sensible heat (and water vapour). In the layer just above the molecular boundary layer, thermal convection is still constrained by the presence of a lower boundary and the dominant type of fluid motion is microscale turbulence induced mechanically by flow over irregularities in the underlying surface. Upward transport is via microscale eddies. The surface or friction layer sustains strong vertical wind shear under conditions of strong upward heat flux; the temperature gradient or lapse rate may be superadiabatic with perhaps a decrease of 5°C in the first few metres. Higher up the microscale turbulence becomes weaker and occurs in short bursts. This intermittency reflects an influence of larger-scale motions, which modulate distributions of static stability and moisture (Munn, 1966; Stull, 1988).

The heat contained in the air or the soil is determined by its temperature and capacity to hold heat. The specific heat (C_g) of a substance is the amount of heat required to raise the temperature of unit mass by unit degree; C_g has units of J kg^{-1} K^{-1}. The volumetric heat capacity (C_v) of a substance is the amount of heat required to raise unit volume by unit degree and has units of Jm^{-3}K^{-1}. Thus $C_g \varrho = C_v$ where ϱ is the density.

The rate of heat flux into the soil is determined by the temperature gradient and the thermal conductivity ($\varkappa$) which is defined as the quantity of heat (J) flowing in unit time (s) through a unit area (m^2) cross section of soil in response to a temperature gradient of 1°C m^{-1} of depth.

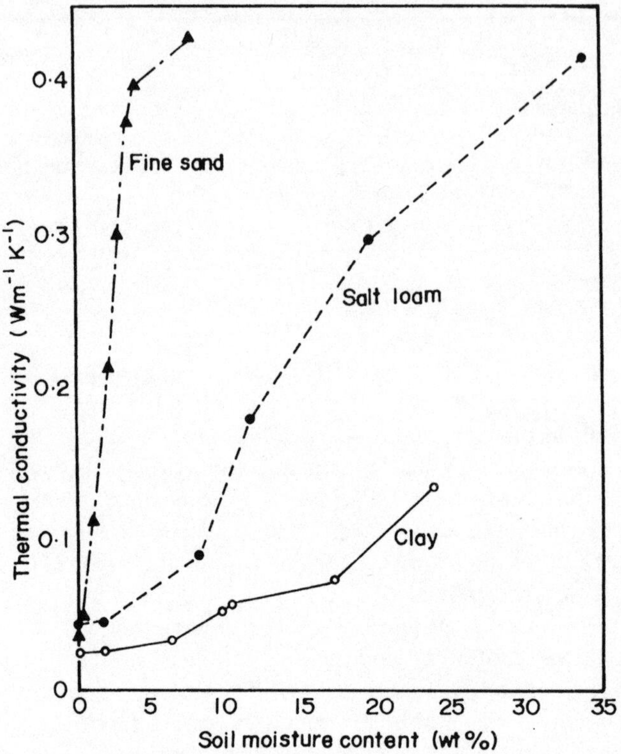

Figure 2.8 Thermal conductivity for three soils as a function of soil moisture content (after Al-Nakshabandi and Kohnke, 1965)

Table 2.5 Thermal properties of natural materials

Material	Density ϱ (kg m^{-3} 10^3)	Specific heat C_g (Jkg^{-1}K^{-1}10^3)	Thermal conductivity κ (W m^{-1} K^{-1})	Thermal diffusivity α (m^{2}s^{-1} 10^{-6})
Clay dry	1.6	0.89	0.25	0.18
Clay saturated	2.0	1.55	1.58	0.51
Sand dry	1.6	0.8	0.3	0.24
Sand saturated	2.0	1.48	2.20	0.74
Still air	0.0012	1.01	0.025	20.5
Turbulent air	0.0012	1.01	~125	~10^7

Source: After Oke, 1978.

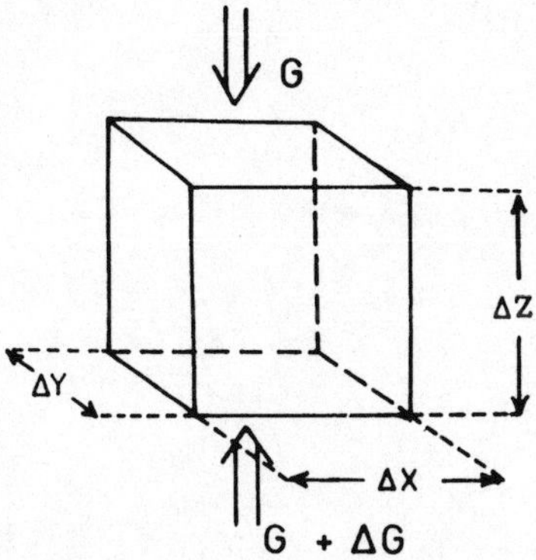

Figure 2.9 Cube of soil $\Delta x\ \Delta y\ \Delta z$ with mean temperature T

Thermal conductivity ($\varkappa$) has units of W m^{-1} K^{-1} and is an empirically determined proportionality factor in the equation

$$G = -\varkappa \frac{\partial T}{\partial z}; \tag{2.21}$$

$\varkappa$ is a function of the composition, moisture content and temperature of the soil; typical values are shown in Figure 2.8 and Table 2.5.

By convection, downward heat flows into the soil are positive and the depth z increases positively downward from the air–soil interface at $z = 0$.

Thermal conductivity determines the rate of heat transfer whereas the thermal diffusivity (α) determines the temperature wave penetration into the soil, and is given by

$$\alpha = \frac{\varkappa}{C_g \varrho} = \frac{\varkappa}{C_v}.$$

As can be seen from Table 2.5, the thermal diffusivity of soil is small — considerably less than that of air at rest. Surface soil temperature changes are observed as waves during the course of a day but the amplitude of the wave will diminish with depth below the surface. The actual temperature profile in the soil is determined by a heat balance, which can be seen by considering a cube of volume V and mean temperature T as shown in Figure 2.9. Assuming no heat flows in the x or y directions, the net heat gain due to heat transfer in time Δt equals the gain minus the loss in Δt. This is

$$\text{net gain of heat} = -\Delta G\ \Delta x\ \Delta y\ \Delta t.$$

This heat warms or cools the volume V, effectively adding an amount of heat $V\Delta(C_v T)$. Therefore

$$-\Delta G\ \Delta x\ \Delta y\ \Delta t = V\Delta(C_v T)$$

That is, with $V = \Delta x\ \Delta y\ \Delta z$

$$\Delta G\ \Delta x\ \Delta y\ \Delta t = -\Delta x\ \Delta y\ \Delta z\ \Delta(C_v T)$$

$$\therefore \frac{\Delta G}{\Delta z} = -\frac{\Delta(C_v T)}{\Delta t}$$

and in the limit as $\Delta t \rightarrow 0$ and $\Delta z \rightarrow 0$

$$\text{i.e. } \frac{\partial G}{\partial z} = -\frac{\partial(C_v T)}{\partial t}$$

$$\text{but } G = -\varkappa \frac{\partial T}{\partial z}$$

$$\therefore \frac{\partial}{\partial z}\left(-\varkappa \frac{\partial T}{\partial z}\right) = -\frac{\partial(C_v T)}{\partial t}$$

Rearranging,

$$\frac{\partial T}{\partial t} = \frac{\partial}{\partial z}\left(\frac{\varkappa}{C_v}\frac{\partial T}{\partial z}\right)$$

$$\frac{\partial T}{\partial t} = \frac{\partial}{\partial z}\left(\alpha \frac{\partial T}{\partial z}\right) \tag{2.22}$$

Solution of this Laplacian equation for constant α shows that the periodic diurnal temperature wave of surface temperature descends downward with a decreased amplitude and an increased phase lag. Experimental results have, however, indicated some discrepancies with this, illustrating that perhaps α is not a constant. This is hardly surprising in view of the variations with depth in soil composition, compaction and moisture found in normal soils. Figures 2.10 and 2.11 illustrate the general effects that are observed. The night-time cool region propagates downward to be lost in the lower depths in late evening. Similarly the marked midday warming of one day is seen as a warmer region at about 30 cm depth on the following day.

Exploiting the theoretical solution of the equation we see that the range of temperature at depth z is

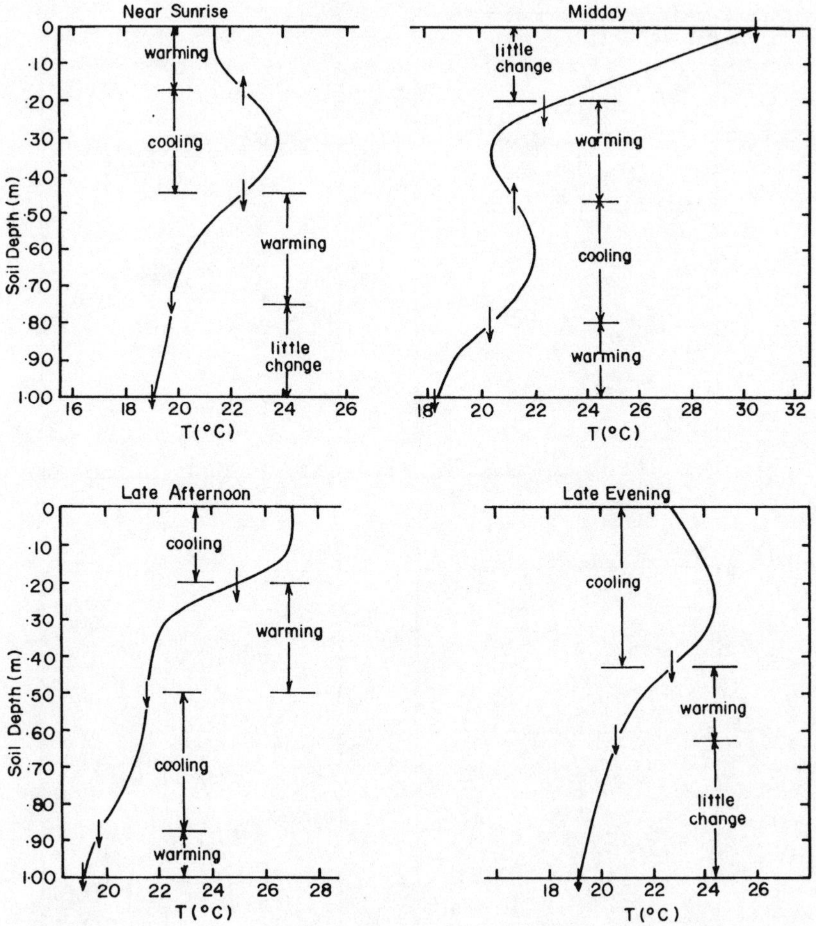

Figure 2.10 Typical variation of temperature with depth at different times of day in summer. Based on data given by Carson (1961) for a yellow-gray silt loam soil near Chicago, Illinois.

$$R_z = R_s \exp\left(-z\left(\frac{\pi}{\alpha p}\right)^{1/2}\right) \tag{2.23}$$

provided that soil properties including porosity, water content and organic matter content are uniform in depth. Here R_z and R_s are the temperature ranges at depth z and the surface respectively, α is the thermal diffusivity and p is the period of oscillation in seconds. Experimental observations tend to confirm equation 2.23 and suggest that temperature ranges are

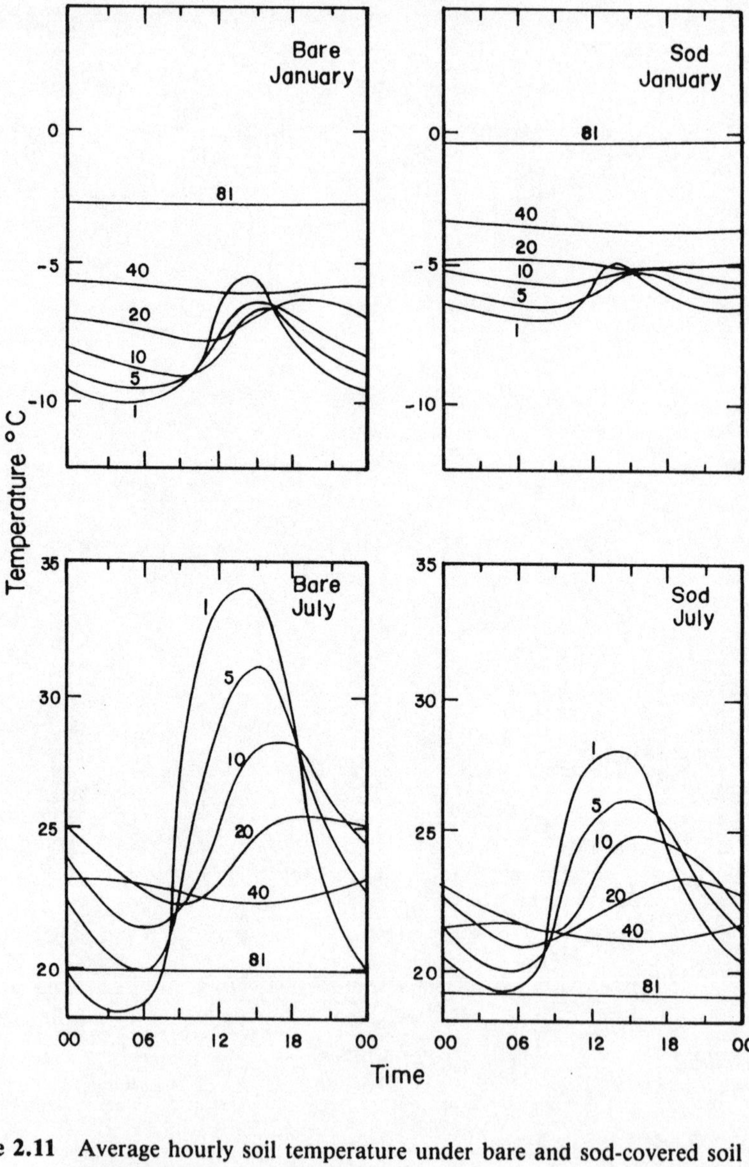

Figure 2.11 Average hourly soil temperature under bare and sod-covered soil at St Paul, Minnesota in January (top) and July (bottom) 1961. Soil depth is shown in cm (after Baker, 1965)

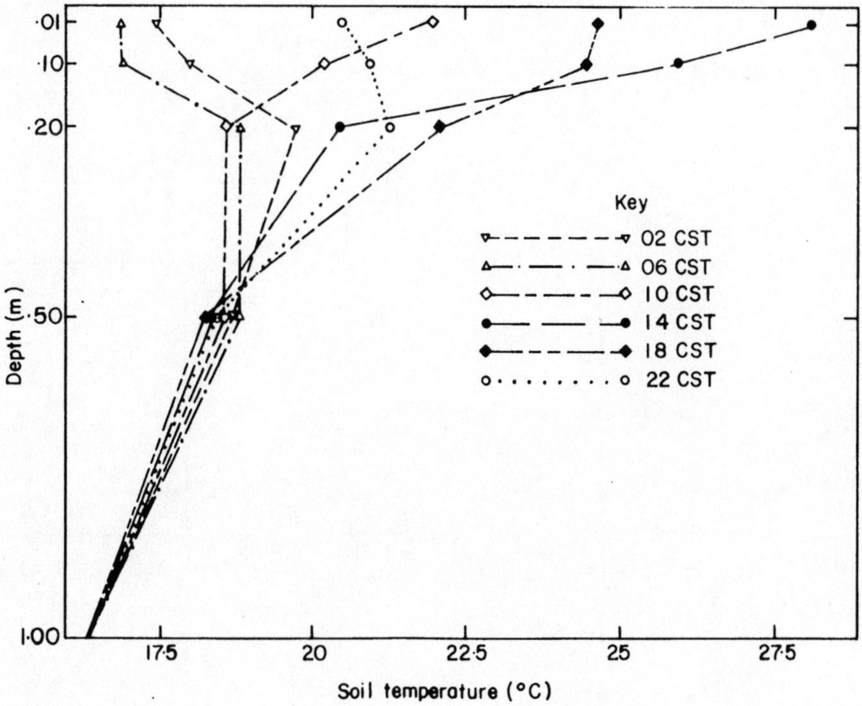

Figure 2.12 Vertical temperature profiles in soil during the course of a typical summer day at Argonne, Illinois, 27 July 1955 (after Carson and Moses, 1962)

approximated by the exponential form. These results suggest that if one digs deep enough, a depth is eventually reached where the temperature fluctuation is insignificant; cellars are an ideal place to store wine. This depth will obviously depend on the soil characteristics but a typical example is shown in Figure 2.12.

Although the results shown in Figure 2.12 are for a diurnal temperature wave at the surface, similar effects are observed for the seasonal variation in surface temperature. Heat is continually moving into and out of the soil and thermal energy is being continually redistributed in the soil as a result of the ever-changing temperature.

Returning to equation 2.20, it is now possible to evaluate R_n, the net

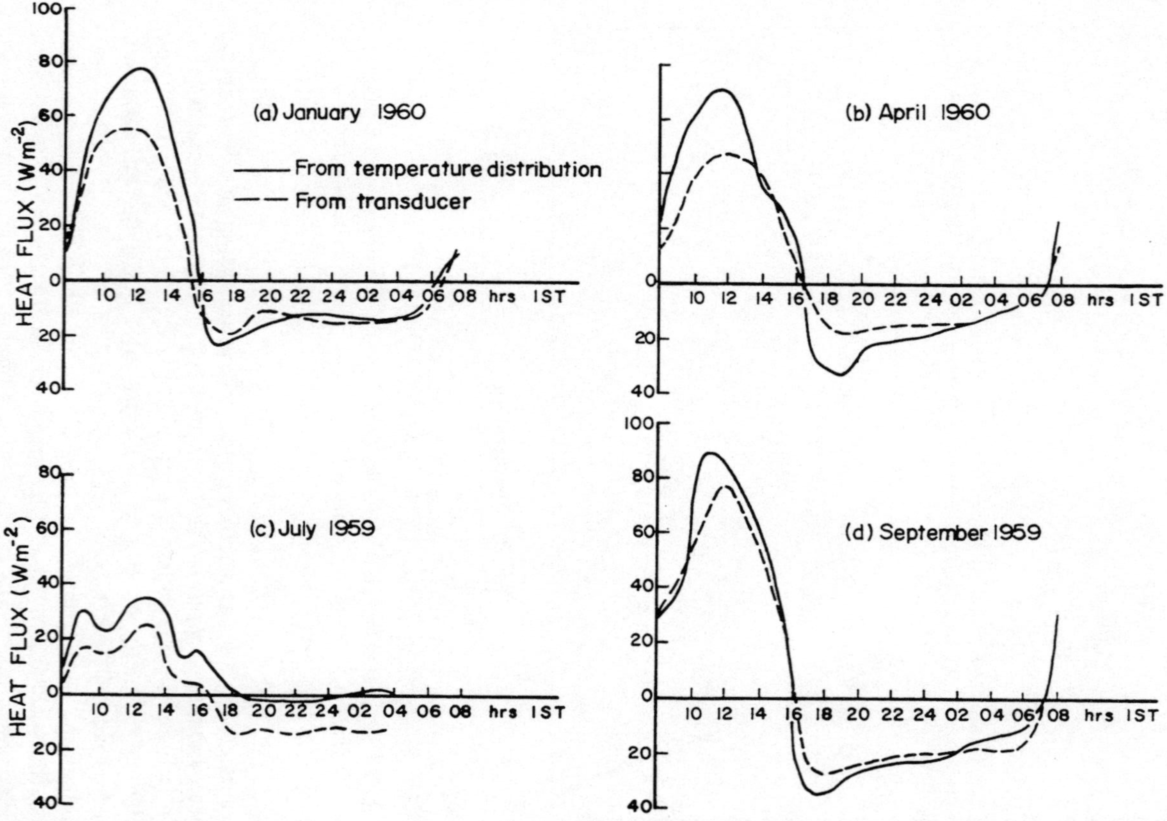

Figure 2.13 Daily cycle of soil heat flux in four different months at Waltair, India (after Padmanbhamurty and Subrahmanyam, 1961)

all-wave radiation received by the surface, and G, the heat supplied by the soil, and infer values of the sum $(H + LE)$.

Unfortunately, one of the major problems of micrometeorology is the determination of the partition of energy between H and LE, knowing their sum. Holtslag and van Ulden (1983) developed an equation expressing H in terms of R_n and its success showed that H is strongly correlated to R_n, as has been observed by other workers. LE on the other hand is strongly dependent on the availability of surface moisture. As moist soil dries out, a larger fraction of the absorbed energy is used to heat the air. A knowledge of soil moisture flux should in theory provide estimates of evaporation and hence latent heat exchange. However, because of horizontal moisture irregularities, horizontal moisture divergence is important and the simple steady-state surface model is invalid (Mahrt, 1987; Wetzel and Chang, 1988).

2.4 Air stability

The surface boundary layer of the atmosphere responds more quickly to temperature changes at the air–soil interface than does the soil. This is because the soil temperature changes primarily by conduction processes, i.e. the molecular motions of the molecules that determine the temperature are transmitted by simple direct coupling or molecular conduction. As we have seen, the atmosphere may gain heat by radiation as, for example, the formation of a radiation inversion due to heat loss to space during the night. The heat is transferred to the air from the soil in a manner we have simulated. Latent heat transfer can also result in cooling at the earth's surface, but a similar warming occurs when the heat is again released during cloud formation.

Also, the atmosphere has motion, and air at different temperatures can simply shift into closer proximity, greatly enhancing the effective heat transfer. If the air moves as a whole, it is considered to advect or undergo advection. If the motions are small eddies or, say, cloud puffs, the motions are convective. Generally the presence of small eddies is highly random and called 'turbulence', whereas the presence of larger convective cells is termed 'convection'. Since the convection rapidly breaks up into smaller eddies, there is little distinction between the two terms. The convection itself can be forced or created by flow over obstacles or rough surfaces. It may be free or formed of instabilities from density or buoyancy differences within the fluid. It may be natural, created by heated surfaces, possibly without the presence of wind.

Presently we will consider the general concept of stability. In the atmosphere it is generally not possible to separate the different types of convection. Often free convection is enhanced by forced motions and vice versa. That is, the air may be unstable for small vertical motions, a motion or perturbation once started grows because it can feed on a buoyancy difference. Free or natural convection may also create thermals or discontinuous jets of warmer air that feed on the surface layers and create motions.

Consider a rising air parcel isolated from its surroundings. As it rises its pressure (p), volume (V) and temperature (T) follow the ideal gas law

$$pV = nRT, \tag{2.24}$$

where n is the number of moles of air and R is the universal gas constant. Alternate forms of this equation are

$$p = \varrho R_a T, \tag{2.25}$$

where ϱ is the density of air and R_a is the gas constant for air which is equal to $R/29$, and

$$p\alpha = R_a T, \tag{2.26}$$

where α is the specific volume. Note that T in all these expressions is in absolute units, K.

Since no heat is allowed to enter or leave the parcel of air as it rises, the first law of thermodynamics becomes

$$\delta q = \delta E + \delta w = 0 \tag{2.27}$$

Here δq is the heat added by some external process, δE is the change in internal energy of the parcel, given simply by $C_v\ \mathrm{d}T$ (C_v is the specific heat at constant volume) and, δw is the work done by the parcel. In this case it is pressure-volume work, corresponding to the volume increase as the parcel rises. In essence the work done by the parcel in expanding is extracted from the internal energy content or sensible heat of the parcel. That is, as the parcel expands, it uses up some of its internal energy to do work on the surrounding environment.

Now $\mathrm{d}(p\alpha) = p\mathrm{d}\alpha + \alpha\mathrm{d}p$.
From equation 2.26, $\mathrm{d}(p\alpha) = R_a\mathrm{d}T$, since R_a is a constant.
Hence,

$$\delta w = p\mathrm{d}\alpha = R_a\ \mathrm{d}T - \alpha\mathrm{d}p.$$

Substituting into equation 2.27, we see that

$$C_v\ \mathrm{d}T + R_a\mathrm{d}T - \alpha\mathrm{d}p = 0.$$

The equation of state for an ideal gas is defined in terms of pressure, temperature and volume. Since the change in internal energy of the air is dependent on the change in temperature, it makes sense to define a specific heat at constant volume (C_v) and constant pressure (C_p). It can be shown that these specific heats are related by

$$C_p = C_v + R_a.$$

Hence, substituting into the previous equation,

$$C_p \, dT - \alpha dp = 0, \tag{2.28}$$

and since

$$\alpha = R_a T/p$$

we find that

$$\frac{dT}{T} = \frac{R_a}{C_p} \frac{dp}{p}. \tag{2.29}$$

Integrating this equation between an initial state T_0, p_0 and final state T, p leads to

$$\frac{T}{T_0} = \left(\frac{p}{p_0}\right)^{0.286}. \tag{2.30}$$

Since $R_a = 287 \text{ J kg}^{-1}\text{K}^{-1}$ and $C_p = 1{,}004 \text{ J kg}^{-1}\text{K}^{-1}$, $R_a/C_p = 0.286$. That is, the temperature decreases as the pressure falls during the ascent of the parcel of air. If we consider the equation for the change of pressure with height, the hydrostatic equation,

$$\frac{dp}{dz} = - \varrho g, \tag{2.31}$$

we see that an alternate form of equation 2.28 is

$$C_p \, dT + g dz = 0 \tag{2.32}$$

$$\text{or} \quad - \frac{dT}{dz} = \frac{g}{C_p} = \frac{9.81 \text{ m s}^{-2}}{1004 \text{ J kg}^{-1} \text{ K}^{-1}}$$

$$\frac{g}{C_p} = \Gamma_d = 0.0098^\circ \text{ K m}^{-1}$$

or 9.8° K km^{-1}.

This is the dry adiabatic lapse rate. The rate of temperature decrease in the atmosphere wherein there is no heat added to or removed from a parcel of air. If a parcel is raised dry adiabatically, without condensation or the addition or removal of heat by an external process, it will cool at a rate of 9.8°K for every 1,000 m of rise or roughly 1°C for every 100 m of rise.

Alternatively, if a parcel at some pressure (p) in the atmosphere is brought down to ground at a standard pressure of 1,000 hPa, it heats so that its temperature at 1,000 hPa is given by:

$$\theta = T\left(\frac{1{,}000}{p}\right)^{0.286} \tag{2.33}$$

This temperature is called the potential temperature of the parcel. If a parcel is subjected only to adiabatic dry transformations as it moves through the atmosphere (that is no heat is either added to or removed from the parcel), its potential temperature, θ, remains constant and is said to be conserved.

A parcel of air will rise through the atmosphere if acted on by a buoyant force. This buoyant force, F, on an air parcel is equal to the weight of the displaced air volume, W_A, minus the weight of the air parcel, W_p.

$$\begin{aligned} F &= W_A - W_p \\ &= g\,V\,(\varrho_A - \varrho_p), \end{aligned}$$

where positive F indicates upward buoyancy, g is the acceleration due to gravity and V is the volume of the air parcel.

From Newton's second law of motion we know that F = ma (mass acceleration) and hence the resulting acceleration experienced by the air parcel due to the buoyant force is

$$a = \frac{F}{m}$$

$$= \frac{g\,V\,(\varrho_A - \varrho_p)}{V\,\varrho_p}$$

$$a = \frac{g\,(\varrho_A - \varrho_p)}{\varrho_p}.$$

Since the pressure experienced by the parcel and the atmosphere are nearly the same, this can be rewritten in terms of the temperature using the equation of state,

$$a = \frac{g(T_p - T_A)}{T_A}. \tag{2.34}$$

The air parcel is conceived of as acquiring buoyancy by changing temperature dry adiabatically as it rises

$$T_p = T_0 - \Gamma_d \, \Delta z,$$

where T_0 is the initial temperature at some level and Δz is the change in height. Taking

$$\Gamma = -\frac{dT}{dz}$$

as the actual lapse rate for the ambient air, equation 2.34 can be rewritten as

$$a = \frac{g \, (\Gamma - \Gamma_d) \, \Delta z}{T_A}.$$

From this it is normal to define a stability parameter S

$$S = \frac{\Gamma_d - \Gamma}{T}, \tag{2.35}$$

and we expect that when S is greater than zero the buoyancy force is downward, and vertical motions are suppressed. Then the air is said to be stable. Similarly, when S is less than zero we expect a positive buoyancy force and the atmosphere is said to be unstable. In the case where S is zero, the atmosphere is classified as neutral or simply dry adiabatic.

In the surface boundary layer the potential temperature can be calculated with sufficient accuracy from the equation:

$$\theta = T + \Gamma_d z. \tag{2.36}$$

This means that

$$\frac{d\theta}{dz} = \frac{dT}{dz} + \Gamma_d = \Gamma_d - \Gamma.$$

$$S = \frac{\Gamma_d - \Gamma}{T} \approx \frac{\frac{d\theta}{dz}}{T} \approx \frac{1}{\theta}\frac{d\theta}{dz}$$

This, in fact, is the standard form of the stability parameter S. The former equation (2.35) really only suffices to explain its physical meaning. The last approximation is valid simply because the absolute temperature has only an insignificant variance within the boundary layer:

$$S = \frac{1}{\theta}\frac{d\theta}{dz}. \tag{2.37}$$

Indeed, it can also be shown that this parameter is proportional to the sensible heat flux and so as far as heat transfer is concerned it is most correct to express all temperatures in terms of potential temperature. This makes sense, for if θ were constant throughout the layer, it would mean that no heat exchanges were taking place, that is, the processes were adiabatic.

Vertical motions in the atmosphere are generally limited by inversions, regions of very stable air in which the temperature actually increases with height. Inversions correspond to a negative lapse rate and are the opposite of lapse conditions in which the normal, negative temperature gradient prevails. We have already considered the formation of the radiation inversion. Other inversions are the advection inversion and the subsistence inversion. An advection inversion is formed when warm air blows across a cooler surface such as a lake or ocean. If the moisture content of the air is sufficiently high, the cooling may form fog. As such an inversion moves inland the base of the inversion rises with increasing distance from the shore.

A subsistence inversion is associated with slowly descending air aloft in high pressure situations with consequent adiabatic warming. This inversion is also observed when air is forced to flow over mountain ranges and may persist for quite long periods.

The air temperature as recorded at a position close to the earth describes a sine curve with the minimum normally occurring in the early morning around sunrise and the maximum occurring sometime after the peak net radiation. These temperatures, in turn, have annual oscillations superimposed, giving a double oscillatory effect.

Figure 2.14 shows temperature gradient contours at different times of the day and different seasons. We see that maximum inversion conditions occur in summer just before sunrise corresponding to maximum surface cooling under clear skies. The annual variation in the time of sunrise and sunset and its effect on atmospheric stability can also be seen.

The daily development of the temperature profile in the surface layer is shown in Figure 2.15. Here we see the surface radiation inversion breaking up at about 10.00 hours in the morning and the development of a most unstable surface layer by 16.00 hours. Such a layer, with a lapse rate much greater than the adiabatic value, is termed super adiabatic and corresponds to strong turbulent mixing. With the approach of sunset, the surface of the earth cools and a stable lapse rate is re-established.

We should remember that humidity is a factor in determining the profile. Though it is normally only 1–2 per cent of the total volume of air, a large amount of heat is retained in water vapour through its latent heat.

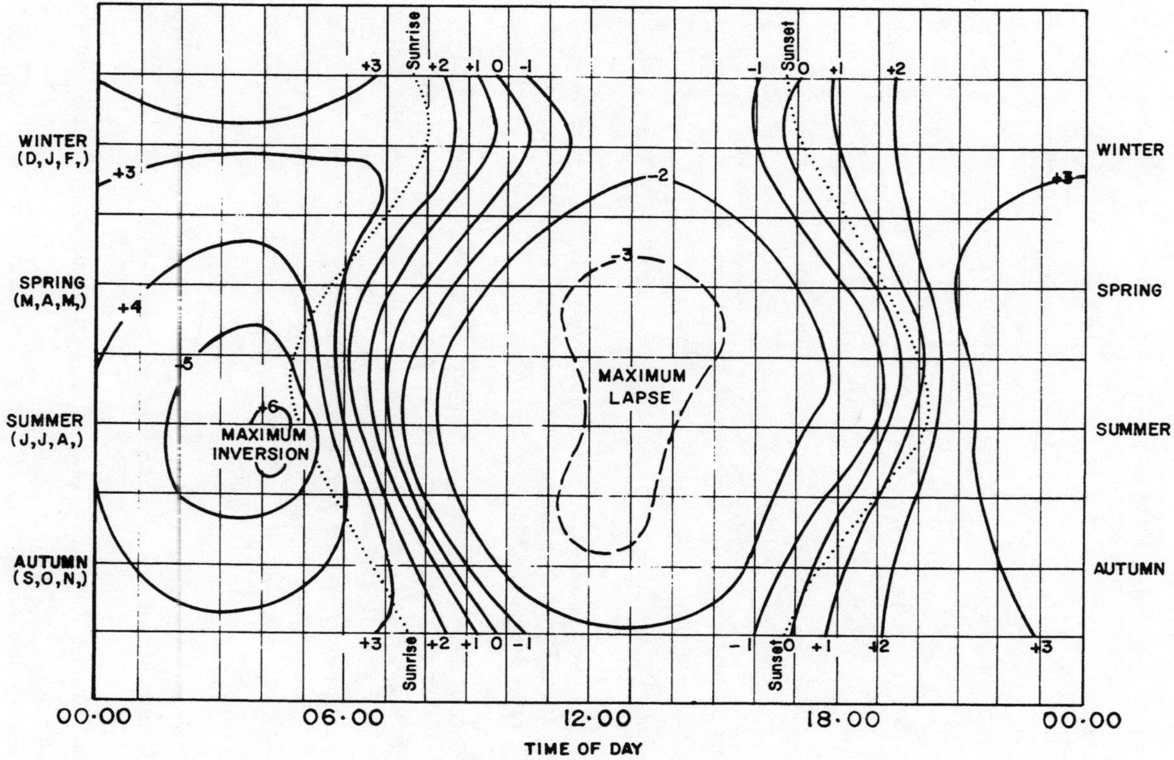

Figure 2.14 Seasonal variation in the vertical temperature difference (°C) between 1.5 and 120 m for each hour of the day (after De Marrais and Islitzer, 1960).

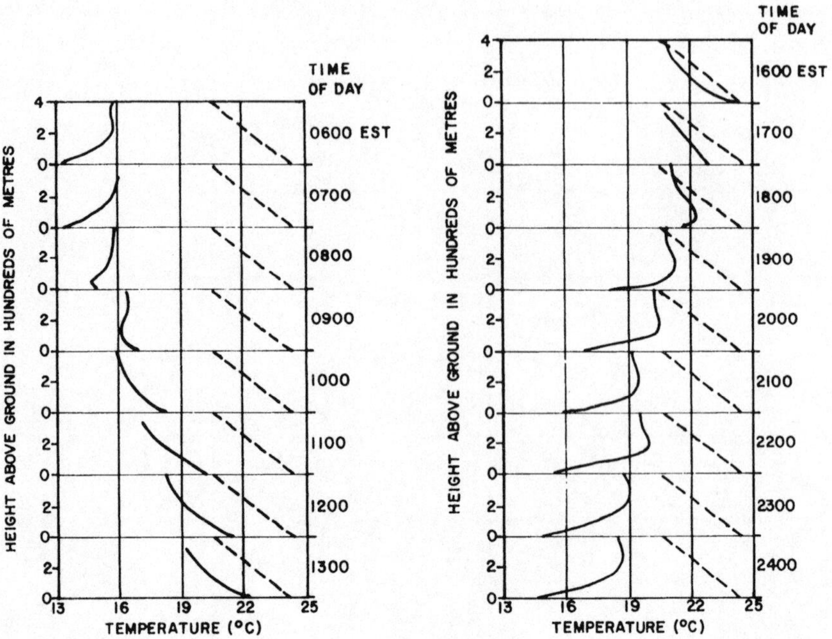

Figure 2.15 The average diurnal variation of the vertical temperature structure. The dashed line in each panel represents the dry adiabatic lapse rate (after Holland, 1953).

Condensation or cloud formation completely alters the stability criteria and the water vapour is transported in much the same fashion as sensible heat.

The other major factor in establishing a strong temperature gradient in the lower atmosphere is the wind. Under light wind conditions it is possible to have marked temperature gradients, particularly at night (see Figure 2.16). As the wind speed increases, mechanical mixing of the air prevents the establishment of either strongly stable or unstable conditions and the lapse rate approaches the dry adiabatic lapse rate (Slade, 1968).

2.5 Local wind structure

The heat and water vapour fluxes are dependent upon the wind and its inherent turbulence. When a fluid (such as air) flows along a surface, fluid parcels close to the surface are slowed down because of viscous forces. Parcels in contact with the surface adhere to it and have virtually zero velocity. Parcels above these tend to slow down due to interactions with the surface parcels and this causes shearing forces. When the flow is slow and

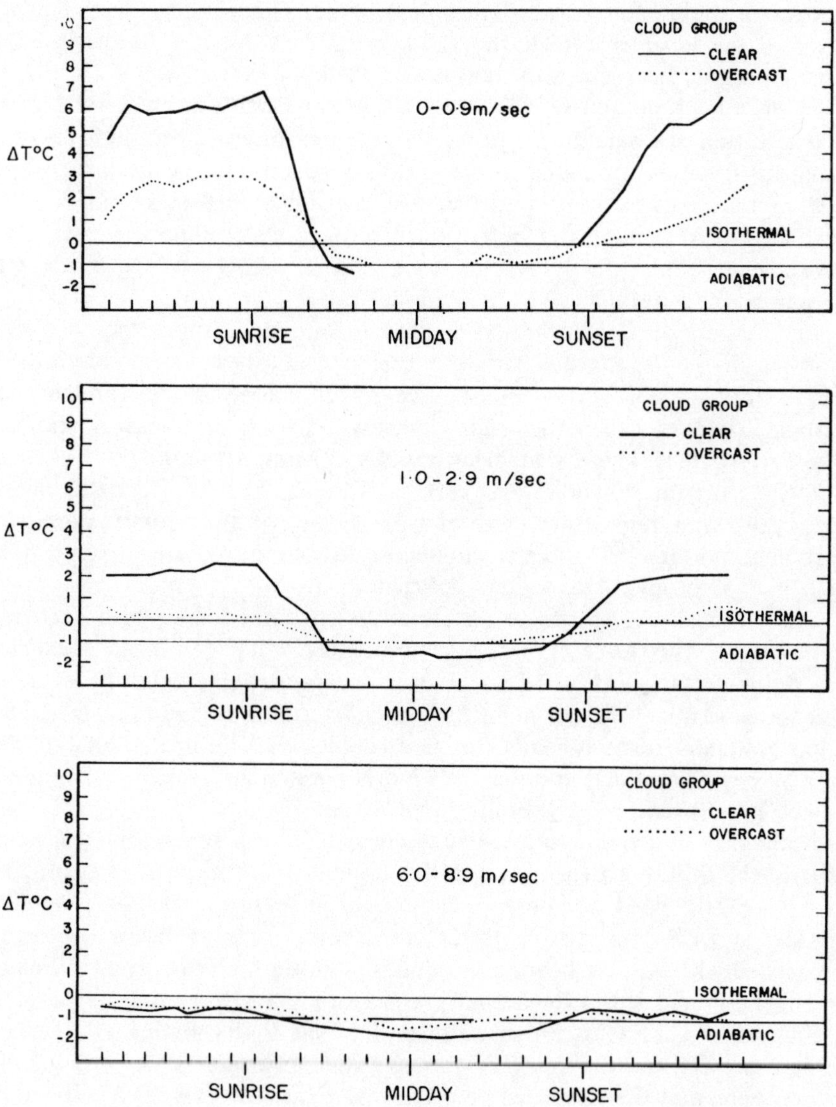

Figure 2.16 Diurnal variation of the temperature difference between 11 m and 110 m as a function of cloud cover and wind speed at 11 m. After Singer and Raynor, 1957). Similar results have been found in Western Australia (e.g. Lyons and Steedman, 1981).

laminar, the interaction, the 'viscous shear', takes place on a submicroscopic level. In faster, turbulent flow the interaction is between lumps or 'globs' of fluid and is called turbulent shear. Specifically, the boundary layer is the layer in which the fluid is affected by the shearing forces originating at the surface (Munn, 1966; Pielke, 1984).

In the first thousand or so metres above the earth's surface wind speed and direction are determined primarily by three forces: horizontal pressure gradient, Coriolis force due to the earth's rotation and the frictional force due to the nearness of the surface. As we move further away from the surface, the frictional force decreases to zero, at which time we observe the geostrophic wind which represents a balance between the horizontal pressure gradient force and the Coriolis force.

Retardation of flow near the surface is caused by molecular viscosity, whereas in the free atmosphere the role played by molecular viscosity is small. Instead, there is a viscous-like effect caused by the presence of turbulent eddies, called the 'eddy viscosity'. This is an apparent viscosity due to turbulent whirls that bring rapidly flowing air from the free fluid and transport the slowly moving surface air aloft. The term 'eddy' is purely descriptive and refers in a general way to any of the infinite variety of turbulent motions that transfer momentum or other properties from regions rich in that property to regions deficient in it.

Eddy viscosity gives rise to an effective frictional force which decreases with height. This force causes a turning of the wind with height such that air at the surface is directed away from the geostrophic wind at an angle; this angle is about 15° at night over rough terrain. Above the surface the wind gradually turns towards the geostrophic wind until at roughly 1,000 m above the ground, the observed wind is parallel and equal in magnitude to the geostrophic wind (Munn, 1966; Slade, 1968).

Eddy viscosity, unlike molecular viscosity, is not a permanent characteristic of the fluid but depends upon the amount of turbulent mixing present. It can be associated not only with mechanical features, such as roughness of the surface, but also with other factors, such as buoyancy forces associated with surface heating or cooling. During the day turbulent mixing is enhanced and the influence of the surface may be detected at a great height. At night the stable stratification of air limits mixing and surface effects are limited to the very lowest layers. Hence, some level of the atmosphere may be subjected to frictional retardation by day and not at night. Above such a level, the average wind speed may be expected to be higher at night than by day. Below, higher speeds will occur during the day, when the faster moving air aloft is mixed downward towards the surface (see Figures 2.17 and 2.18).

Night-time records also illustrate a wind-speed maximum somewhere in the middle of the planetary boundary layer (e.g. Kamst and Lyons, 1982a). This low-level jet is thought to be due largely to the 'decoupling' of the

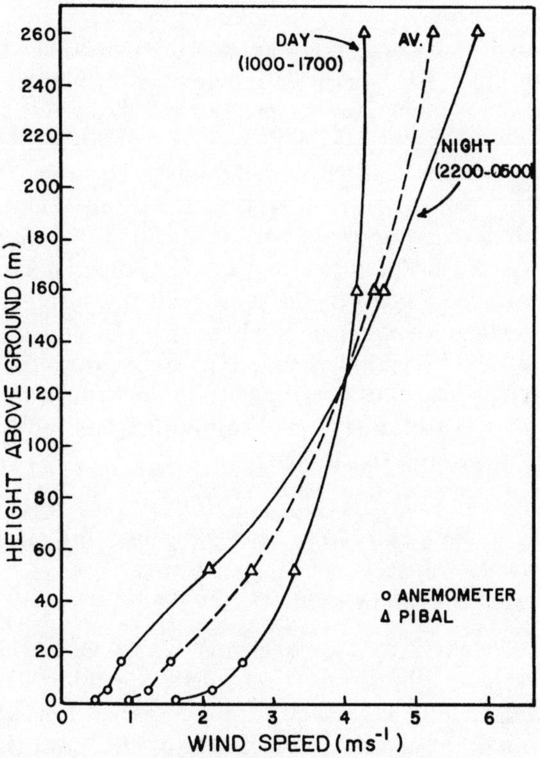

Figure 2.17 Average wind-speed profiles constructed from measurements made during a one-year period (after Holland, 1953)

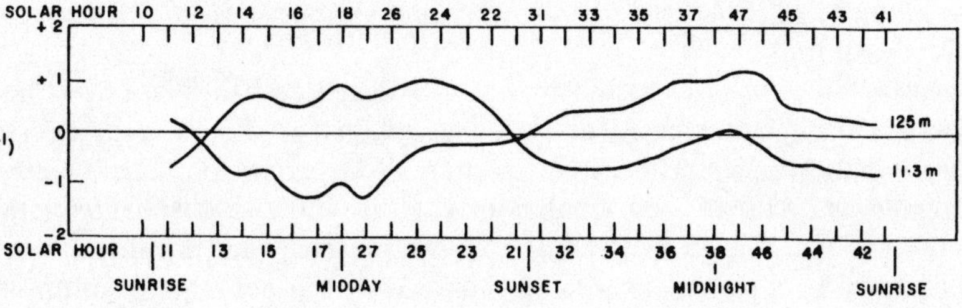

Figure 2.18 The deviation of the mean hourly wind speeds from their respective daily averages at 11.3 m and 125 m (after Singer and Raynor, 1957)

upper layers of the planetary boundary from the surface and the setting up of an inertial oscillation.

Wind speed and direction may change with height in the absence of turbulent mixing due to large-scale pressure distributions. Hence, the vertical variation of wind cannot be specified totally by this simple picture.

Interaction of weather systems on all scales results in a three-dimensional wind speed and direction that varies continuously in time. This turbulence is an atmospheric characteristic that causes the diffusion of pollutant. Analysis of such a condition consists of dividing the fluctuating wind observed at a point into a mean motion and superimposed fluctuating motions. The distinction between the mean and turbulent motion may be determined on the basis of the dimensions of the diffusion system, as in the size of a puff of smoke. Wind fluctuations larger than the puff tend to move it in its entirety and thus contribute to the mean motion. Fluctuations considerably smaller than the puff tend to pull it apart and thus contribute to the turbulent dispersion of the puff.

Turbulence is dependent upon three factors:

(i) mechanical effects of objects protruding into the air stream,
(ii) the vertical rate of increase of wind speed, and
(iii) the vertical temperature structure.

Mechanically generated turbulence is greater with higher wind speeds and since it is generated at the surface it characteristically exhibits a decrease with height. If the mean wind vector were constant with height, turbulence would consist entirely of a small displacement of the mean flow by surface configurations. However, wind shear means that if a particle of air is displaced from one level to another, it arrives at the new level with some of its initial momentum and will constitute a perturbation in its new surrounds. Any vertical fluctuations existing in a fluid with vertical shear will result in a chaotic field of vertical and horizontal turbulence. Vertical shear initiates or enhances turbulence when the shear is great (Slade, 1968; Stull, 1988).

An upward flux is expected to be transferred by turbulence so that the properties being transferred are propagated in the direction of decreasing values of these properties, that is down the gradient. The greater the turbulence or gradient, the greater the rate at which this transfer or mixing can take place. Injection of smoke in the presence of turbulence will lead to the down gradient transport of smoke with the consequent dilution and spreading of the smoke mass.

We should reiterate that the division between mean and turbulent motion depends on the scale of averaging and the nature of the phenomenon being studied. Also, although both transport and turbulence are of importance in horizontal transfer, turbulence is the main mechanism for vertical transfer.

The concept of a random process has been applied with considerable success to a statistical description of atmospheric turbulence. In general, atmospheric turbulence is described as being stationary, homogeneous, isotropic and Gaussian.

'Stationary' means that its statistical properties derived for a time series do not vary with time. Obviously major changes, such as the diurnal cycle or the passage of a front, mitigate against stationarity. The concept is generally fulfilled during periods of several hours duration when the large-scale weather pattern is not changing.

'Homogeneous' means that the statistical properties will not vary in space. Horizontal homogeneity is usually expected in regions with little differences in topography. Owing to the vertical forces of gravity and buoyancy and the physical boundary of the surface, vertical homogeneity is a rarity in the real atmosphere.

'Isotropic' means that the statistical properties of the field are independent of the rotation of the coordinate axis. This condition is met in a limited way in the atmosphere, but is not the general case.

'Gaussian' means that the frequency distribution of values follows the familiar Gaussian or normal curve. This distribution is completely specified by the mean and standard deviation. No turbulence is completely Gaussian but turbulence is so difficult to measure more thoroughly than these statistics allow that the other statistics of the motion are often ignored.

Specifically, the mean of a particular wind statistic is defined as

$$\overline{\mathrm{W}} = \frac{1}{T}\int_0^T W\mathrm{d}\tau, \tag{2.38}$$

where T is the sampling time. Using the standard orthogonal coordinate system, the wind components at any instant are represented as,

$$\begin{aligned} u &= \overline{u} + u' \\ v &= \overline{v} + v' \\ w &= \overline{w} + w' \end{aligned} \tag{2.39}$$

where $\overline{u}$ is the mean component in the x direction (normally east → west), $\overline{v}$ is the mean component in the y direction (north → south), $\overline{w}$ is the mean component in the z direction (vertical), and u', v', and w' are the fluctuations about the respective means.

The intensity of turbulence along each axis is defined as

$$i_x = \left(\frac{\overline{u'^2}}{\overline{u}^2}\right)^{1/2} = \frac{\sigma_u}{\overline{u}}$$

$$i_y = \left(\frac{\overline{v'^2}}{\overline{v}^2}\right)^{1/2} = \frac{\sigma_v}{\overline{v}}$$

$$i_z = \left(\frac{\overline{w'^2}}{\overline{w}^2}\right)^{1/2} = \frac{\sigma_w}{\overline{w}} \tag{2.40}$$

where σ is the standard deviation of the velocity distribution. Another statistic used is the variance, or standard deviation, of the azimuthal wind direction angle (σ_θ^2 or σ_θ), where θ is the angular direction of the horizontal wind (Slade, 1968).

Taking the x-axis as being parallel to the mean horizontal wind direction, then $\overline{v} = 0$ and hence

$$v = v' \text{ (since } \overline{v} = 0)$$
$$u = \overline{u} + u'$$

$$\tan \theta' \approx \theta' = \frac{v'}{\overline{u} + u'}$$

$$\overline{\theta'^2} = \overline{\frac{v'^2}{(\overline{u} + u')^2}} = \overline{\frac{v'^2}{(\overline{u})^2} \left| 1 + \frac{u'}{\overline{u}} \right|^{-2}}.$$

Expanding the denominator and neglecting higher order terms,

$$\overline{\theta'^2} = \frac{\overline{v'^2}}{(\overline{u})^2}, \tag{2.41}$$

that is

$$\sigma_\theta = \frac{\sigma_v}{\overline{u}}. \tag{2.42}$$

One should note that i_x, i_y, i_z and σ_θ are not constants for a particular set of meteorological conditions, but rather their values depend on the sampling and averaging times that are inherent characteristics of the data set.

A typical time trace of turbulence is shown on Figure 2.19. The trace may be regarded as Eulerian, with the wind direction measured by a wind vane at a particular fixed point. Alternatively, it could represent the path of a parcel following the wind; in this case it represents a Lagrangian trace. The difference between these two frames of reference is illustrated in Figure 2.20. which shows the Eulerian observer (E) fixed at the source, watching the pollutant move downwind, whereas the Lagrangian observer (L) rides on the puff as it is advected. In the Lagrangian frame of reference we are following the motion of an individual parcel of air and the measurements are taken relative to that parcel. For example, a free-floating neutrally

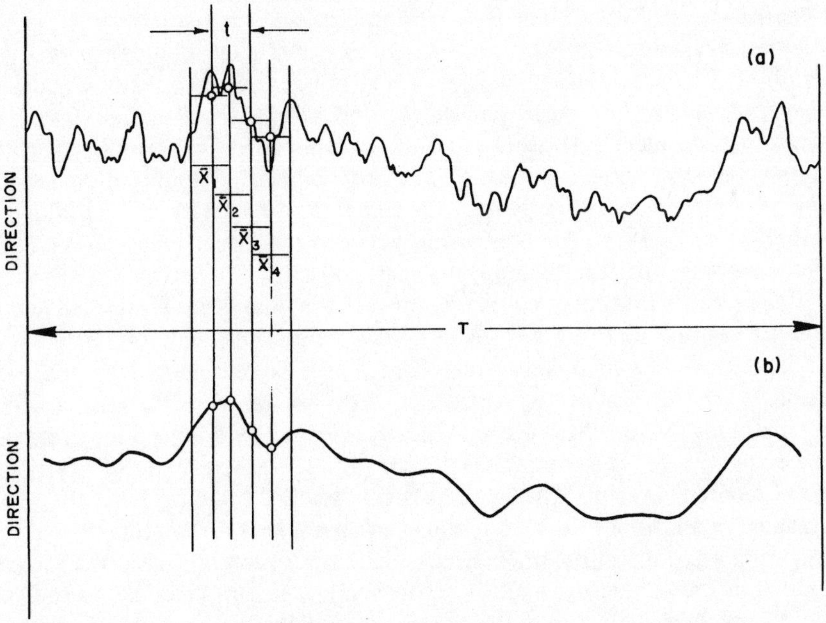

Figure 2.19 An instantaneous wind-direction trace (a) and the time-averaged trace derived therefrom (b). The time interval *t* represents the averaging time, and *T* represents the total length of record, or the sampling time (after Slade, 1968).

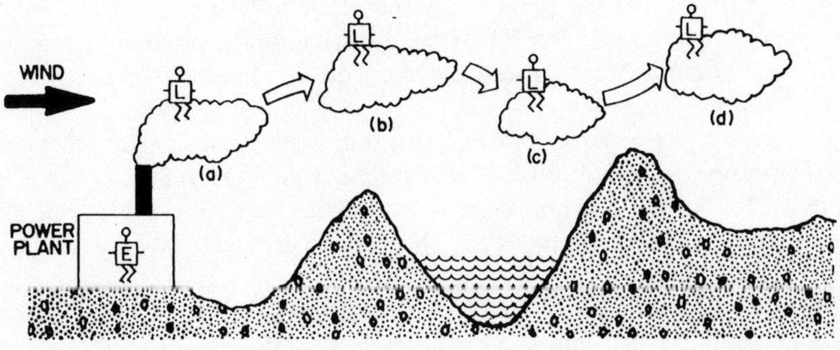

Figure 2.20 Comparison between Eulerian, E, and Lagrangian, L, observers (after Drake *et al.*, 1979). Copyright © 1979 Electric Power Research Institute. Reprinted with permission.

buoyant balloon-borne sensor would return Lagrangian observations. (A balloon is neutrally buoyant if it has no tendency to rise or fall.) On the other hand, if observations are taken relative to a fixed point then they are not observations of a single parcel of air, but rather observations made of many parcels as they pass that point, and are in the Eulerian frame of reference. Turbulent diffusion of pollutant is more easily formulated in the Lagrangian sense, whereas most of our observations of pollutant concentration or meteorological variables are Eulerian measurements. Although a theoretical basis exists for converting between the two frames of reference it is extremely difficult to apply to real flows.

Usually measurements are made on a trace and samples are taken as discrete, digital values to give an Eulerian series. The sampling time, T, is the total time in which data are recorded and is represented by the total period of record. The averaging time, t, is the smaller time interval in which the continuously varying trace may be represented by a constant value; this time is set by the electronics, the inertia of the vane, or/and the response of the recorder system. The whole system should be set up so that the set averaging time removes fluctuations of periods shorter than t. When measuring wind-direction fluctuations, increasing sampling time leads to an increase in the standard deviation up to some point; as more measurements are taken, more of the long-period fluctuations are sensed. Increasing averaging time decreases the standard deviation as shorter period fluctuations are damped out.

Atmospheric diffusion is related not only to the value of the standard deviation but also to the frequency or range of frequencies in the spectrum that makes the greatest contribution to the total standard deviation. Two records may have the same standard deviation but one may be caused by a few oscillations of long period and the other by more numerous oscillations of shorter period. The long-period oscillations tend to transport the puff of pollutant whereas the shorter-period fluctuations tend to tear it apart.

Experimental data have shown that the spectra of stable air parcels exhibit little variance and most of it is mechanically generated. Atmospheric instability leads to a thermal component in the spectrum and the shifting of the spectra to lower frequencies. This effect can be clearly seen in Figure 2.21.

The standard deviation of wind-direction fluctuations responds to the effects of stability as shown on Figures 2.22 and 2.23. Vertical motions exhibit less deviation in heavy winds than in light winds during unstable conditions. Under stable conditions the reverse situation holds at low levels with forced convection causing greater fluctuations.

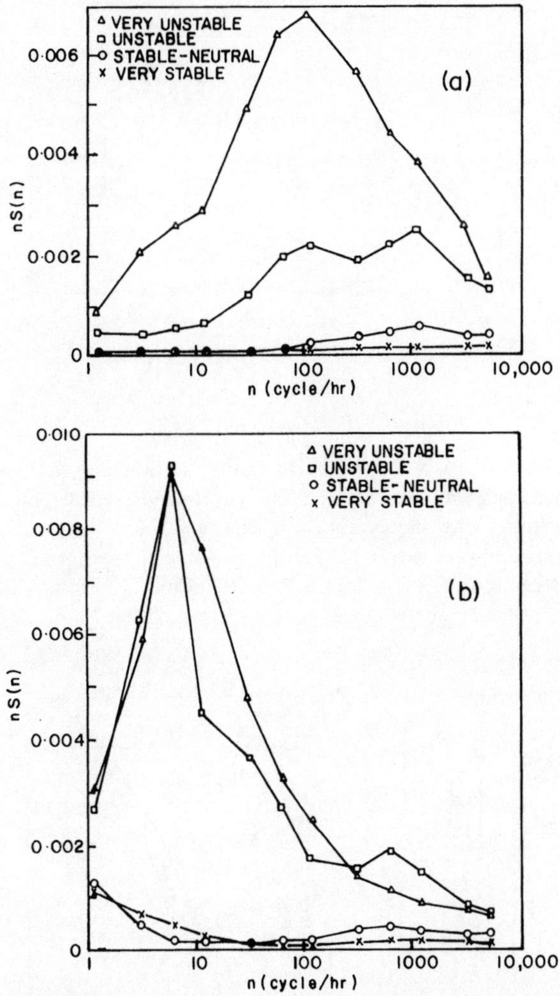

Figure 2.21 Spectra of the vertical (a) and horizontal (b) wind-direction fluctuations at 15 metres grouped by atmospheric stability (after Walker, 1963)

2.6 The logarithmic profile

The atmospheric boundary layer is normally modelled as a fully developed wind profile over homogeneous and uniformly rough terrain. The overriding geostrophic wind is considered to have been constant for a sufficiently long time that a steady-state condition exists in the wind profile. This presumes effectively that an infinite plane with such properties exists upwind of the measurement area; that is, the measurement point has an

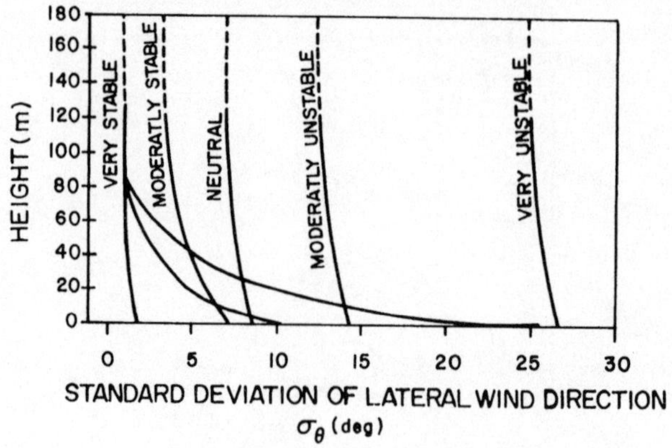

Figure 2.22 The vertical variation of the lateral wind-direction standard deviation for various stability regimes. The curves represent average or typical conditions with the exception of the two outer 'very stable' lines, which represent extremes. Sampling times of the data used in the construction of this diagram averaged about 10 minutes and averaging times were on the order of a few seconds (after Slade, 1968).

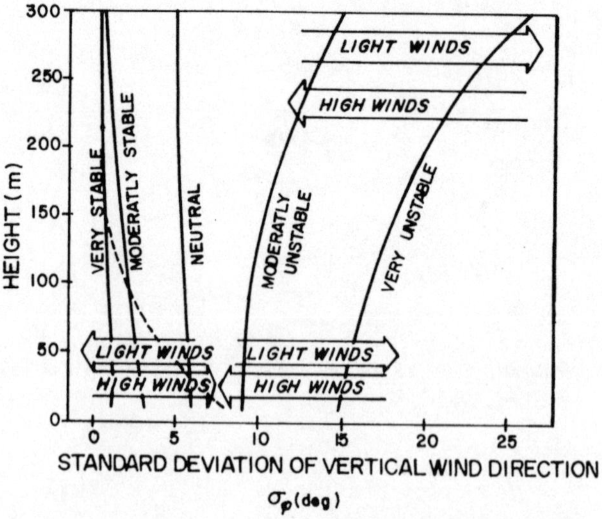

Figure 2.23 The vertical variation of the vertical wind-direction standard deviation for various stability regimes. The relations shown represent sampling times of at least 3 minutes at the lowest levels and up to 30 minutes at the greatest heights (after Slade, 1968).

'infinite fetch'. Such a fetch requirement ensures that experiments are not contaminated by local topographic variations and that results are reproducible. Such an infinite fetch, of course, never occurs in atmospheric flows. This is a basic limitation of not being able to control the vagaries of atmospheric motion; we cannot take a slice of the atmosphere into the laboratory for experimentation.

To relate the boundary layer to usual engineering approaches it is instructive to consider the turbulence experiment conducted by Osborne Reynolds. He experimented with flows in pipes and found that dynamic similarity obtains in the flow provided the dimensionless group

$$\mathrm{Re} = \frac{uz}{\nu} \tag{2.43}$$

is equivalent. That is, provided the product of the fluid velocity, u, and a characteristic dimension, z, divided by the kinematic viscosity, ν, is the same, flows have the same features. This number is called the Reynolds number and is used to scale experimental results. It is an accurate specification of the characteristics of flow in pipes or simple situations that have a well-defined characteristic length, z.

In particular, in pipe flow, it has been found that the extremely smooth characteristics or 'lamina' of flows of low velocity, 'laminar flows', break down into turbulent flow when the Reynolds number is about 2,000. However, the identification of a characteristic length in the atmospheric boundary layer is not self-evident. The natural length determined by the pipe diameter of Reynolds experiments is not present and even the presence of a wing chord, as would appear on an airfoil, is absent. Even the defined scale created by the wire mesh or grid in a wind tunnel is absent.

Since typical atmospheric velocities are of the order of m s^{-1}, ν is 0.15 $cm^2\ s^{-1}$, and distances are typically in hundreds of metres, it is obvious that any Reynolds number calculated for the atmosphere is large ($>$ 100,000). Consequently, atmospheric flows are normally turbulent. In fact, the effective Reynolds numbers are so large, that the Reynolds number criteria has no real significance in atmospheric science. Normally, the concept of smooth flows over smooth boundaries is of no relevance; we consider only turbulent flows over rough boundaries.

We consider an atmospheric wind profile given by the mean wind, $\overline{u}$, as a function of height, z, above the ground. The drag of the solid earth retards the wind so that it approaches zero at $z = 0$. The force of retardation per unit horizontal area is known as the shearing stress, τ (Newton's m^{-2}). The wind shear, the change of the wind velocity with height, is a consequence of the presence of this shearing stress.

In the case of laminar flow, it is known that the shearing stress is proportional to the velocity gradient. The proportionality factor is μ, the dynamic viscosity. Thus,

$$\tau = \mu \frac{d\bar{u}}{dz}. \tag{2.44}$$

The dynamic viscosity is related to the kinematic viscosity by the relation:

$$\nu = \frac{\mu}{\varrho}$$

At 20°C, μ and ν for air are 1.81×10^{-5} kg m^{-1} s^{-1} and 1.5×10^{-5} m^2 s^{-1} respectively.

Newton's equation and molecular viscosity account adequately for the transfer of fluid properties very near boundaries and, in general, in any small volume of fluid. It results from the transfer through molecular exchange due to the random motions of molecules.

In a turbulent situation, however, the transfer is through the turnover of eddies. The existence of shear can generally be equated with a transport of momentum. Eddy motions that move air upward decrease the momentum of the air at the higher level; on the other hand, downward movements of air increase the momentum of the lower layers. In any case, on the average there is a net transport of momentum from the higher levels to the ground.

By analogy with molecular transport, it is usually assumed that eddy transport can be characterised in much the same way as molecular transport. The concept, in fact, can be derived from mixing length concepts such as conceived by Prandtl and summarised in the last paragraph. Here we will simply rely on the analogy and write:

$$\tau = (\mu + A) \frac{d\bar{u}}{dz}, \tag{2.45}$$

where A is termed the exchange coefficient. This form takes care of both molecular and eddy processes with the respective coefficients. Continuing the analogy, we define an eddy–viscosity coefficient for momentum transport, K_m, as

$$A = \varrho K_m, \tag{2.46}$$

where K_m corresponds to the molecular kinematic viscosity. Note that it is possible to define other eddy coefficients for water vapour transport, heat transport, etc., since all these quantities are generally conveyed down the gradient.

Invariably μ is much less than ϱK_m, especially in the atmosphere, so equation 2.45 becomes

$$\tau \approx \varrho K_m \frac{\mathrm{d}\bar{u}}{\mathrm{d}z}. \tag{2.47}$$

In general, the shearing stress τ is expected to vary with height in the lowest layers of the atmosphere. However, it is found experimentally that the first 20 to 200 metres of the atmosphere exhibit a characteristic constant shearing stress, that is, a layer in which there is little or no change of τ with height. This layer is defined as the 'constant stress' layer or turbulence boundary layer. For this layer, one can define an effective velocity, u_*, the friction velocity, such that

$$u_* = \sqrt{\frac{\tau_0}{\varrho}}. \tag{2.48}$$

where u_* is a characteristic of the surface friction or shear stress, τ_0. To this day there is no satisfactory unified physical model of the wind profile in the constant stress layer. Hence we shall follow the usual procedure and use dimensional analysis to get a general theoretical form to describe the wind in this layer.

Dimensional analysis is a powerful technique that can be used to develop our understanding of any system when we know the controlling parameters but do not have a model through which to relate the individual parameters (Blueman, 1983). For example, consider the right-angle triangle shown in Figure 2.24 made up of sides of length *a, b, c.* The units of area are m^2 and therefore one would expect the area of the triangle to be related to the square of one of the sides times some function of the angle between two of the sides. That is,

$$\text{area of triangle} = a^2\ \mathrm{f}(\Phi).$$

Dimensionally both sides of this equation have the same units (m^2) and we have not specified the function, f. The triangle can be redrawn as two smaller right-angle triangles as shown. Hence following our dimensional argumcnts

$$\text{area of triangle 1} = c^2\ \mathrm{f}(\Phi)$$
$$\text{area of triangle 2} = b^2\ \mathrm{f}(\Phi)$$

but the sum of these two areas is equal to the area of the original triangle, therefore

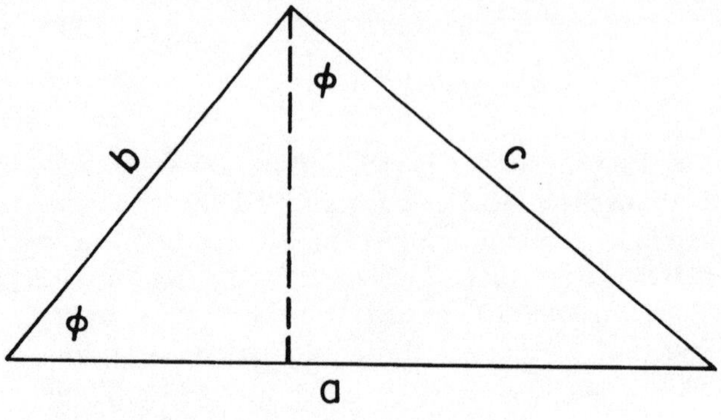

Figure 2.24 Area of triangle for dimensional analysis

$$a^2 f(\Phi) = b^2 f(\Phi) + c^2 f(\Phi),$$

i.e.

$$a^2 = b^2 + c^2.$$

This is the classic Pythagorean result which took Pythagoras much longer to derive. It shows the power of dimensional analysis.

In our case the vertical structure of $\bar{u}$ or the shear $d\bar{u}/dz$ appears to depend on the following quantities: the kinematic viscosity, ν; the height above the surface, z; the air density, ϱ; and the frictional stress, τ. Thus, from dimensional analysis we would expect a relationship between the five quantities,

$$\frac{d\bar{u}}{dz}, \ \nu, \ z, \ \varrho, \text{ and } \tau.$$

Dimensional analysis states that two independent dimensionless ratios can be formed from these five parameters. This is given by the number of quantities (5) minus the number of fundamental dimensions (length, mass, time, i.e. 3). These ratios are:

$$\left(\frac{d\bar{u}}{dz}\,\frac{z}{u_*}\right) \text{ and } \left(\frac{zu_*}{\nu}\right).$$

If these parameters completely describe the system then one of the dimensionless ratios must be a function of the other, that is

$$\frac{d\bar{u}}{dz}\frac{z}{u_*} = f\left(\frac{zu_*}{\nu}\right)$$

or

$$\frac{d\bar{u}}{dz} = \frac{u_*}{z} f\left(\frac{zu_*}{\nu}\right), \tag{2.49}$$

where f is an unknown function called the universal function.

Within millimetres of the earth's surface the vertical velocity profile must be governed mainly by molecular viscosity, because close to the surface rough elements shield the surface and there can be no vertical motions. That means that turbulence due to eddies becomes negligible, there being insufficient height for the eddies to come into play. That is, close to the surface equation 2.44 holds so

$$\frac{d\bar{u}}{dz} = \frac{\tau}{\mu} = \text{constant} = \frac{u_* z}{\nu} = \text{Re, the Reynolds number.}$$

In fact, except for flow over very smooth ice or still water, we may ignore the effect of the Reynolds number on the vertical profile, and thus

$$\frac{d\bar{u}}{dz} = \frac{u_*}{kz}, \tag{2.50}$$

where the proportionality constant k is known as von Karman's constant; it has been found experimentally to have a value of about 0.4 (Businger *et al.*, 1971; Goddard, 1970; Webb, 1970; Hicks, 1976; Kondo *et al.*, 1978).

Integrating equation 2.50 within the constant stress layer, we find that

$$\bar{u}(z) = \frac{u_*}{k} \ln(z) + \text{const}, \tag{2.51}$$

which is the common 'logarithmic profile', generally observed in wind tunnels. Usually the constant of integration is so defined as to introduce the concept of surface roughness z_0:

$$\text{Const} = -\ln z_0,$$

where z_0 is effectively the height above the plane surface at which $\bar{u} = 0$.

$$\text{Then } \bar{u}(z) = \frac{u_*}{k} \ln\left(\frac{z}{z_0}\right). \tag{2.52}$$

This expression is valid only for $z > z_0$, that is, above the roughness elements, since the dimensional argument only applies in this region. Note the z_0 effectively scales the profile or the height is measured in 'roughness lengths'. Typical values of z_0 are given by Wieringa (1980), Kondo and Yamazawa (1986) and Stull (1988). These illustrate that higher surface obstructions are associated with larger aerodynamic roughness lengths, but in all cases z_0 is smaller than the physical height of the surface element.

In addition, it is usually necessary to take account of the possibility that the actual zero-plane datum level used in an experiment may differ from the effective zero-plane position. This results either from an error in our measurement of the vertical or, simply, from our inability to specify the ground surface in this simple model.

It is easiest to envision this, for instance, with the presence of long grass or corn, in which the wind does not penetrate to the ground. The appropriate expression is:

$$\bar{u}(z) = \frac{u_*}{k} \ln \left(\frac{z - d}{z_0} \right), \tag{2.53}$$

where d is the zero-plane displacement.

The existence of the logarithmic profile near the surface of the earth in purely mechanical turbulence has been attested by numerous experiments. Still, the simple logarithmic wind profile fits real winds only under neutral conditions, with a large fetch, and then only applies to, perhaps, the first 100 metres. Truly, it is strictly applicable only to wind-tunnel studies of shear flow and is of limited use in the atmosphere. Generally, wind profiles show different curvatures at night than during the day. This observation led Deacon (1949) to propose the empirical expression:

$$\frac{d\bar{u}}{dz} \propto z^{-\beta}, \tag{2.54}$$

where the logarithmic profile obtains when $\beta = 1$ and the profile is linear on a logarithmic scale. Otherwise, β is greater than unity for a stable, night-time situation and less than unity for a super adiabatic lapse rate such as would occur in the afternoon (Oke, 1978). The profiles of temperature and humidity show similar curvature trends (Rider, 1954).

Effectively the logarithmic profile results only from mechanical driven turbulence with a dry adiabatic, neutral-lapse condition. As already illustrated, this condition occurs only just after dawn and at dusk. At other times the profile deviates from this neutral condition and exhibits either an ability to enhance vertical motion, and is unstable, or an ability to suppress vertical motion, and is stable. This is due to buoyancy and results from a flow of heat to or from the underlying surface by conduction, convection and radiation.

In the unstable situation the vertical motions feed on the mechanically driven eddies and grow, producing convection and turbulence. In the stable situation, however, turbulence can only occur when the stable, buoyancy forces are overcome by the mechanically driven eddies. In this case Richardson postulated that turbulence should occur in the atmosphere when the production of turbulent energy is just large enough to compensate for the energy absorbed (consumed) by the stable, buoyancy forces. That is, the ratio:

$$\frac{\text{Rate of consumption of turbulent energy by buoyancy forces}}{\text{Rate of production of turbulent energy by wind shear}}$$

should give a measure of the onset of instability. Substituting parameters, Richardson found the dimensionless group:

$$\mathrm{Ri} = \frac{g \dfrac{\mathrm{d}\theta}{\mathrm{d}z}}{T_a \left(\dfrac{\mathrm{d}\bar{u}}{\mathrm{d}z}\right)^2},$$

which is normally derived from data for two height levels, z_1 and z_2, using the mean absolute temperature, T_a, for mean height, z_a, defined by

$$z_a = \left(\frac{z_1^2 + z_2^2}{2}\right)^{1/2}, \text{ harmonic mean;}$$

or

$$z_a = (z_1 \, z_2)^{1/2}, \text{ geometric mean.}$$

An alternative form for the Richardson number is:

$$\mathrm{Ri} = \frac{g(\Gamma_\mathrm{d} - \Gamma)}{T_a \left(\dfrac{\mathrm{d}\bar{u}}{\mathrm{d}z}\right)^2},$$

which shows its direct relationship to the departure from the adiabatic lapse rate. The Richardson number is negative for unstable conditions and positive for stable conditions.

In the usual case the lapse rate is not adiabatic but, rather, there is an exchange of heat between a moving parcel and its environment. That is, the process is diabatic. In diabatic shear flow, the eddies transport heat as well as momentum because the temperature of the eddies differs from that of the surrounding air. Hence the eddies transport both heat and momentum. By analogy with the momentum flux an eddy-heat transport flux can be written as

$$H = \varrho\, C_p\, K_H\, (\Gamma - \Gamma_d)$$

where C_p is the heat capacity of the air at constant pressure and K_H is the coefficient of eddy-heat conductivity. Hence

$$\mathrm{Ri} = \frac{gH\, K_m}{C_p \tau T_a \left(\dfrac{d\bar{u}}{dz}\right)^2 K_H},$$

since

$$\tau = \varrho K_M \frac{d\bar{u}}{dz}.$$

This leads to an alternative definition of the Richardson number, the flux Richardson number,

$$\mathrm{Rf} = \frac{\mathrm{Ri}\, K_H}{K_M} = \frac{gH}{C_p\, T_a \tau \left(\dfrac{d\bar{u}}{dz}\right)}.$$

Both the Richardson number and the flux Richardson number may be used to characterise the stability of the boundary layer. However, they are not constant in the layer and depend on the height difference over which they are measured.

Obviously a parameter that is dependent on the height difference over which it is measured will not provide a unique function to incorporate the effects of buoyancy in the simple wind profile. We have assumed uniform, straight, parallel turbulent flow near the surface with constant stress and heat flux. It has also been assumed that this flow will extend to some depth into the atmosphere, wherein the assumptions of constant heat and momentum will be valid.

It can be shown from the governing equation of motion that the effect of thermal buoyancy enters this problem through a buoyancy parameter, g/T_A. Constant momentum and heat flux likewise lead to dependence of the flux on the dimensional parameters u_* and $H/C_p\varrho$ respectively.

These parameters characterise the velocity and temperature profiles in the surface layer, and Monin and Obukhov (1954) used them to define the unique length scale,

$$L = \left(\frac{u_*^3}{K\ (g/T_A)}\right)\left(\frac{-H}{C_p\varrho}\right)^{-1}$$

$$= \left(\frac{u_*^3\ C_p\ \varrho\ T_A}{K\ g\ H}\right). \tag{2.55}$$

This length scale is a constant, characteristic length scale for any particular example of the flow. It is negative in unstable conditions (upward heat flux), positive for stable conditions and approaches infinity as Γ approaches Γ_d. Unfortunately a physical interpretation of L is difficult though its magnitude is thought to be related to the size of the largest energy-containing eddies and roughly corresponds to the height at which the shear and buoyancy production (or destruction under stable conditions) rates of turbulent kinetic energy are equal (Wyngaard, 1988). Its tendency to approach infinity poses numerical difficulties, so that often the reciprocal $1/L$ is used instead of L.

It is now possible to repeat a dimensional analysis of the parameters responsible for determining the wind profile, using this time the Monin–Obukhov length, L, as the parameter specifying the diabatic influence. This leads to a functional form:

$$\bar{u}(z) = \frac{u_*}{k}\ \mathrm{f}\left(\frac{z}{L}, \frac{z_0}{L}\right) \tag{2.56}$$

or its equivalent form,

$$\bar{u}(z) = \frac{u_*}{k}\left(\mathrm{f}\ (\frac{z}{L}) - \mathrm{f}\ (\frac{z_0}{L})\right), \tag{2.57}$$

since the role of z_0 is merely a constant of integration and hence only shifts the velocity profile without changing its form.

To accord with the simple logarithmic profile, we define a further function, Φ_u, such that:

$$\frac{\partial \bar{u}}{\partial z} = \frac{u_*}{kz}\ \Phi_u\left(\frac{z}{L}\right), \tag{2.58}$$

called the diabatic influence function or non-dimensional wind gradient.

It has been verified by Batchelor (1953) that Φ_u is a function of z/L only, as expected from equation 2.56. Thus the function Φ_u can be expanded as a Taylor series in z/L and, if non-linear terms are ignored,

$$\Phi_u = 1 + \beta\left(\frac{z}{L}\right),$$

which is the simple log-linear profile first derived by Monin and Obukhov; the constant β is derived by experiment. While the form was initially proposed only for conditions close to neutral, it can give good results in stable conditions for a range of stabilities $0 < z/L < 1$ with β ranging between 4 and 7 (McVehil, 1964; Webb, 1970; Sheppard *et al.*, 1972; Kondo *et al.*, 1978).

Further, assuming that the eddy diffusion coefficient for momentum and heat transfer are equal, the so-called 'OKEYPS' interpolation equation

$$\phi_u^4 - \gamma\left(\frac{z}{L}\right)\phi_u^3 = 1$$

can be derived. It is strictly applicable to unstable conditions with a value of γ of about 18 (Lumley and Panofsky, 1964). Although this formulation gives a reasonable fit for unstable profiles, the assumption of equality of eddy diffusion coefficients is questionable (Swinbank and Dyer, 1967; Swinbank, 1968; Dyer and Hicks, 1970; Businger *et al.*, 1971) and the general form itself does not easily permit its use. Nevertheless, the diabatic influence function allows us to account for the deviations from the simple logarithmic profile that are observed under varying stabilities.

So far in our discussion of atmospheric motion we have derived a wind-speed profile with an increase with height above the surface as a result of a net downward transport of momentum by turbulent eddies. In fact the mean wind does not increase indefinitely with height and it has been observed that the effect of turbulence decreases with elevation and is usually negligible above several thousand metres. Moreover, the eddy stress has been found by analysis of wind fluctuation observations to decrease with height above the surface layer.

None the less, the assumption of constant stress implies that both turbulence and the mean wind will increase with height. Consequently, this model applies only in the lowest part of the atmosphere and outside this layer we must incorporate the possibility of diminished stress, the effects of the pressure gradient and the Coriolis force.

2.7 The Ekman spiral

A force/momentum balance on the air in the horizontal plane gives the following relationships (see, for example, Sutton, 1953):

$$\varrho \frac{d\bar{u}}{dt} - \varrho f \bar{v} = - \frac{\partial p}{\partial x} + \frac{\partial}{\partial z} (\tau_{zx}) \tag{2.59}$$

$$\varrho \frac{d\bar{v}}{dt} + \varrho f \bar{u} = - \frac{\partial p}{\partial y} + \frac{\partial}{\partial z} (\tau_{zy}) \tag{2.60}$$

where ϱ is the air density, p the pressure, τ_{zx} and τ_{zy} are the components of the horizontal shearing stress, and f is the Coriolis parameter (equal to $2\Omega \sin \phi$ where Ω is the angular velocity of rotation of the earth and ϕ is the latitude).

Physically, the situation is simply that the drag of the surface is balanced by momentum drawn from the layer of the atmosphere which forms the planetary boundary layer or the friction layer. This extraction of momentum is apparent not only in the reduction of wind speed as the ground is approached, but also in the turning of the wind from its direction outside the boundary layer. The motions responsible for the vertical transfer of momentum will also diffuse material carried by the air, and the planetary boundary layer may be regarded as one in which material is more or less rapidly mixed and outside which material penetrates only slowly.

If we assume steady (i.e. unaccelerated) flow the terms involving

$$\frac{d}{dt}$$

in the above equations are equal to zero. Following the eddy diffusion expression, we suppose that

$$\tau_{zx} = \varrho K_M \frac{\partial \bar{u}}{\partial z}$$

$$\tau_{zy} = \varrho K_M \frac{\partial \bar{v}}{\partial z}.$$

That is, the eddy viscosity can be generalised by breaking it up into x and y components which, in general, depend on height.

Hence equations 2.59 and 2.60 become

$$\varrho f \bar{v} - \frac{\partial p}{\partial x} + \varrho \frac{\partial}{\partial z} [K_M \frac{\partial \bar{u}}{\partial z}] = 0 \tag{2.61}$$

$$- \varrho f \bar{u} - \frac{\partial p}{\partial y} + \varrho \frac{\partial}{\partial z} [K_M \frac{\partial \bar{v}}{\partial z}] = 0. \tag{2.62}$$

The velocity shear decreases with height and, since the effect of turbulent friction also decreases with elevation, the third terms in these equations, which represent the downward momentum transported by eddy turbulence, should become negligible at some height in the atmosphere. If we orient the x axis in the direction of the wind at this level, $v = 0$ in equation 2.61 and $\partial p/\partial x = 0$. Here the pressure gradient force exactly balances the Coriolis force; the wind flows along the isobars with an intensity given by

$$\overline{u}_g = \frac{-1}{f\varrho} \frac{\partial p}{\partial y}.$$

This is the well-known geostrophic wind equation; and $\overline{u}_g$ is the geostrophic wind.

If we allow that, below the level of the geostrophic wind, the eddy diffusivity is constant and that the pressure gradient is independent of height, the general form of the wind profile that satisfies equations 2.61 and 2.62 is,

$$\overline{u}(z) = \overline{u}_g \, (1 - e^{-az} \cos az) \tag{2.63}$$

$$\overline{v}(z) = \overline{u}_g \, e^{-az} \sin az \tag{2.64}$$

where

$$a = \left(\frac{f}{2\,K_M}\right)^{1/2}.$$

Near the ground, friction causes the air to flow across the isobars in the direction of low pressure. This effect decreases with increasing height and disappears at the level of the geostrophic wind, which can conveniently be defined as the lowest level at which $\overline{v} = 0$ and therefore the lowest level at which the wind is parallel to $\overline{u}_g$. From the above, this must occur when $az = \pi$ and since $f \approx 10^{-4}$ sec^{-1} and K_M is of the order of m^2 s^{-1}, the depth of the layer of frictional influence in the atmosphere is of the order of hundreds of metres.

The solution to the wind distribution in the planetary boundary layer given by equations 2.63 and 2.64 was first obtained by Ekman in 1902 and is known as the Ekman spiral. Although originally derived for ocean layers, it has never been valid in the upper layers of the ocean. Nevertheless, it provides a reasonably good qualitative explanation of the wind structure of the atmosphere despite the use of the constant eddy-viscosity assumption. As we have already noted the eddy viscosity must vary with height in the planetary boundary layer increasing just above the ground as larger eddies become effective and then decreasing at greater elevations as the general

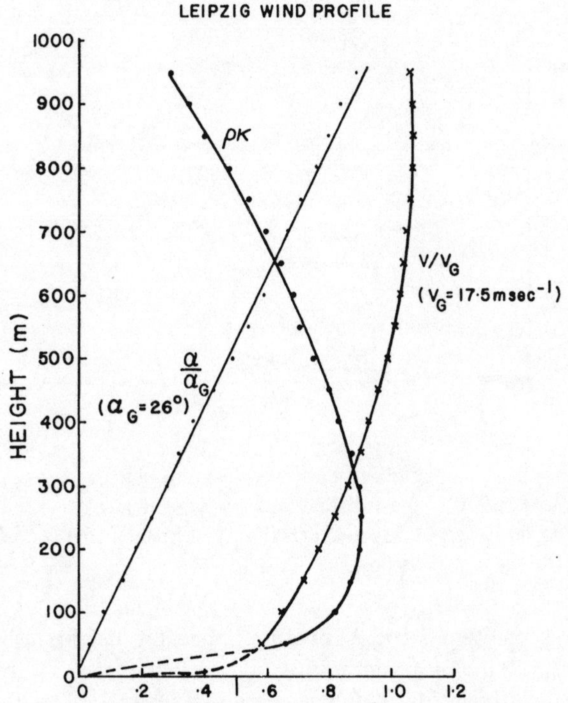

Figure 2.25 The vertical profile of wind and eddy viscosity (V is resultant horizontal wind speed, V_G the geostrophic value, a the change of wind direction from that at the surface, a_G the value corresponding to the geostrophic direction) (after Pasquill, 1974).

influence on the airflow of the surface frictional drag decreases. Moreover, we can expect – on the basis of the analysis of the diabatic surface layer – that the eddy structure in the planetary boundary layer will also be strongly influenced by buoyant heat fluxes. Nevertheless, under steady conditions the Ekman spiral is a realistic description of the wind profile.

Figures 2.25 and 2.26 show the general form of K as a function of height. Figure 2.25 shows the exchange coefficient $A = \varrho K$ together with the relative wind profile u/u_g and relative specific volume α/α_g. Figure 2.26 shows a normalised eddy viscosity, Kf/u^2, as a function of height. Note that K reaches a maximum and then decreases (although at the upper levels the wind shear is so small the actual value of K is unimportant), the third terms – the eddy diffusion terms – in equations 2.61 and 2.62 being insignificant (Pasquill, 1974).

This indicates that an otherwise satisfactory expression of the wind

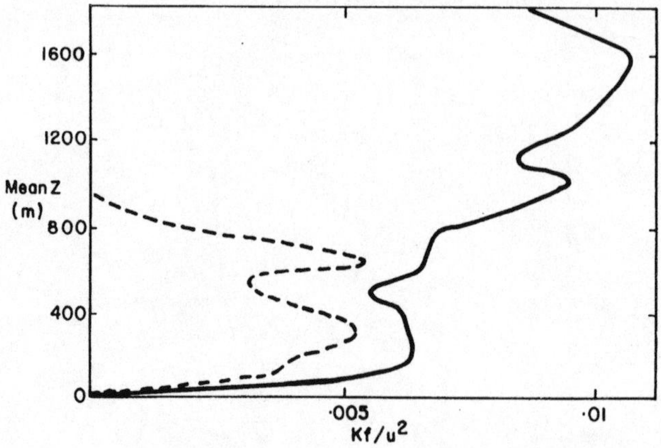

Figure 2.26 Profiles of normalised eddy viscosity in near-neutral conditions (after Clarke 1970). The full line is the profile derived by Clarke from wind-profile observations in Australia. The broken line is the 'Leipzig' profile (after Pasquill, 1974).

profile is not sensitive to the K profile at greater heights. Physically, this stems from the fact that the major part of the momentum transfer is carried out at relatively low heights where the shear and geostrophic departure of the wind are greatest, and once these are correctly represented, the upper part of the wind profile cannot be sensitively modified. More sophisticated numerical analyses also show that the mean fields are relatively insensitive to the parameterization of turbulence (e.g. Huang and Raman, 1989). From the geostrophic wind equation we can calculate the pressure gradient using the magnitude of the geostrophic wind,

$$\frac{\partial p}{\partial x} = -f \varrho v_g,$$

where v_g is the y-axis component of the geostrophic wind. Substituting into equation 2.59, assuming steady conditions, and integrating from the ground ($z = 0$) to height z, leads to an expression for the change of τ;

$$\tau(z) - \tau(0) = f \int \varrho (v_g - \bar{v})\, dz,$$

where the subscripts on τ have been omitted for convenience. With this equation and a vertical profile of the mean wind and its direction, it is possible to calculate $\tau(z)$ at any height. With τ values, a vertical profile of K_m can be found.

If the x-axis is taken along the surface wind direction, v may be neglected. Taking $\tau(0)$ as the resultant stress at the surface and assuming a small enough z, such that the wind has not turned appreciably from its direction at the surface, $\tau(z)$ is the resultant stress at height z. We approximate the height dependent $v_g = V_g \sin \alpha \approx V_g \alpha$ for small differences, where V_g is the geostrophic wind speed and α is the total turning of direction in the friction layer. With these approximations

$$\frac{\tau(0) - \tau(z)}{\tau(0)} \approx \frac{\varrho \, fz \, V_g \, \alpha}{\tau(0)}.$$

Observations over land give values near 10^{-3} for $\tau(0)/\varrho \; V_g^2$, the geostrophic drag coefficient, and 0.3 radians for α. Taking $V_g = 10 \text{ ms}^{-1}$ and $f = 10^{-4} \text{ s}^{-1}$,

$$\frac{\tau(0) - \tau(z)}{\tau(0)} \approx 3 \times 10^{-5} \, z,$$

indicating that the percentage decrease in τ is about 10 per cent at a height of thirty metres. That is, the use of the logarithmic form to describe the profile of the wind is only expected to be valid in the first thirty or so metres of the atmosphere since this model relies on an assumption of constant shear stress. Above this height the wind is described by the Ekman spiral.

2.8 Turbulence

In our discussion of the logarithmic wind profile we have concentrated on the mean wind velocity. However, as already noted, the wind consists of both mean and turbulent motion and we need to understand this turbulence in greater detail to appreciate its role in the diffusion of atmospheric pollutants.

The initial studies of turbulence were carried out by Reynolds and this led to his definition of the Reynolds number. Reynolds showed that for $Re <$ 1,000–2,000 there was laminar motion and for $Re > \approx 20{,}000$ the motion was always turbulent. The arbitrariness of these limits is basically due to the manner in which the experiments were conducted and the inability to remove unwanted irregularities. Nevertheless, the indefinite nature of turbulence appears even more vague in atmospheric motions, as – generally – a length scale cannot be specified precisely and, in any case, all reasonable scales produce enormous Reynolds numbers. Hence it is generally correct to say that the atmosphere is always turbulent. Despite this, the nature of turbulence and its effectiveness in transport and diffusion depend on the relevant time and space scales. Here we will attempt to describe the properties of turbulence and derive some quantitative descriptions to characterise its observed properties.

Turbulence can generally be defined as 'ignorance' in relation to fluid flows. It is seen as a cascade of useful mechanical energy, from large scales to ever-smaller scales, ending up in molecular motions or heat. Lumley and Panofsky (1964) suggest that the basic properties of turbulent motion are rotational, dissipative, three-dimensional, stochastic, non-linear, diffusive and that it is a continuum phenomenon of large time and length scale. Agreement is not universal on all of these properties but they are a starting point.

Turbulence is observed to be rotational and dissipative; it acts to transform mechanical energy into internal energy. There are random three-dimensional motions that are approximately irrotational and non-dissipative, such as a body of water the surface of which is disturbed by a turbulent wind, or a fluid outside a turbulent boundary layer, but these do not dissipate mechanical energy to internal energy through a cascade of eddies of diminishing size. This cascade, ending in dissipation, is bound up with non-linearity and three-dimensionality as well as rotationality, and is generally regarded as an essential feature of turbulence.

Turbulence is three-dimensional. The cascade of energy to smaller and smaller eddies can be thought of as taking place by vortex stretching (described by non-linear terms in the equations of motion), which force the motion to be three-dimensional and non-linear.

Turbulence is stochastic. In practice, no matter how carefully the conditions of an experiment are reproduced, the velocity field cannot be predicted in detail. Although the equations of motion have not been shown to have a unique solution for a given set of conditions, uniqueness of the physical containment seems likely. However, the solution appears to be sensitive to minute changes in the conditions and quite non-unique. We may never know or be able to prescribe the conditions finely enough to predict the detailed structure of the flow. For this reason, and also because we do not want or need information on the details of the flow, we use a statistical description.

Turbulence is non-linear. The transfer of energy from one size eddy to another (vortex stretching) can take place through a production of smaller multiples of the original motion or 'higher harmonics', which are non-additive or non-linear.

Turbulence is diffusive. A tagged point in a turbulent fluid will wander about, making excursions further and further from its initial location, qualitatively like the motion of a molecule in a gas. This behaviour is responsible for the transport of properties such as mass, momentum and heat, and is a far more effective form of transport than that resulting from molecular motion alone. In a turbulent flow, quantities such as heat and mass transfer and drag are generally considerably increased, and it is usually in this way that turbulence makes itself felt.

In turbulent flow, the time and length scales of the turbulent motion are quite large and often of the same order as the time and length scales of the properties being transported. Molecular diffusion is described by a model in which the time and length scales are very much smaller than the scales of the property being diffused. Describing turbulence as being qualitatively similar to molecular diffusion, though possible, must always be approached with caution, for it implies a disparity in scales that is not often present.

In a general way it is fair to say that the mean action of turbulence is not, except under unusual conditions, describable by a differential equation, since local stress (local in terms of the mean distributions) is not described by local conditions. The fluid usually interacts with itself over a wider area than will permit such a description.

Turbulence is a continuum phenomenon. In most flows the smallest dynamically significant length scale is very much larger than intermolecular distances or molecular dimensions.

In discussing atmospheric turbulence it is usually convenient to assume that the turbulence is homogeneous. Homogeneous turbulence occurs when the probability distributions describing the flow are independent of position in the fluid. It is usually assumed that surface roughness elements are randomly distributed and that turbulence is homogeneous in any horizontal plane. The presence of the earth's surface and the associated shear assures that atmospheric turbulence is not homogeneous in the vertical direction.

In analysing the behaviour of this continuous stochastic process we must resort to a statistical description of the phenomenon. Hence the need to define a mean. The mean of any motion may be expressed as:

(i) an Eulerian time mean and/or Eulerian space mean;
(ii) a Lagrangian time mean;
(iii) an ensemble or stochastic mean.

The last (iii) type of mean is our usual way of forming averages, adding-up and dividing by the total number. The stochastic mean can be considered as a statistical average over many realisations or many nominally equivalent experiments.

Normally, we are dealing with Eulerian observations; air sensors such as anemometers, are fixed in space and the Eulerian time mean is defined as,

$$\bar{u} = \begin{cases} \dfrac{1}{n} \displaystyle\sum_{i=1}^{n} u_i & \text{for discrete observations,} \\[2ex] \dfrac{1}{T} \displaystyle\int_{t_1}^{t_2} u \, \mathrm{d}t & \text{for continuous observations.} \end{cases} \tag{2.65}$$

The instantaneous horizontal wind speed, u, is a function of time and space, and can be represented as a sum of the mean velocity, $\bar{u}$, and the fluctuating (turbulent) component, u'. That is,

$$u = \bar{u} + u' \tag{2.39}$$

where u' is the instantaneous fluctuation about the mean. Obviously, the mean of u', by definition is zero. The variance of the horizontal wind speed is defined as

$$\overline{(u')^2} = \sigma^2 = \frac{1}{T}\int (u - \bar{u})^2 \, dt$$

and, if we consider the relationship of u to the vertical wind speed, w, we can define a covariance

$$\overline{u'w'} = \frac{1}{T}\int (u - \bar{u})(w - \bar{w}) \, dt.$$

The mean square of u', or variance of u', is a measure of the intensity of the fluctuation of u; the square root of the variance is the standard deviation. The quantity σ/u is the intensity of turbulence (equation 2.40).

At this point, the equations of mass continuity and motion will be considered; equations containing fluctuating components. In reducing these equations, the statistical quantities take various forms, which can be simplified by using averaging concepts known as the Reynolds' rule of averaging. That is:

1. $\overline{\bar{u}} = \bar{u}$, i.e. the mean of the mean is the mean.

2. $\overline{u'} = \overline{(u - \bar{u})} = \bar{u} - \bar{u} = 0$

3. $\overline{uw} = \overline{(\bar{u} + u')(\bar{w} + w')}$

   $= \overline{\bar{u}\bar{w}} + \overline{\bar{u}w'} + \overline{u'\bar{w}} + \overline{u'w'}$

   $= \bar{u}\,\bar{w} + \overline{u'w'}$

4. $\overline{u^2} = \bar{u}^2 + \overline{u'^2}$

where the overbar implies a mean of the quantity.

It is generally assumed that the turbulence is stationary during an experiment. That is, the probability distribution of the fluctuation is independent of time during the period of the experiment. If this were not true, statistical results from any experiment would depend on the time at which they were taken and it would be impossible to compare data from the same experiment, let alone between experiments. The opposite of stationarity is non-stationarity, which perhaps gives a negative definition of stationarity. Turbulence is non-stationary with respect to the mean if it exhibits trend, that is, a gradual increase or decrease with time. It is non-stationary with respect to the intensity if the turbulence undergoes bursts of varying intensity, and non-stationary with respect to the spectrum if the turbulence experiences bursts of different frequencies; this implies a shift of intensity from one frequency to another.

Non-stationary turbulence implies that the statistical properties of the fluid are dependent on the time of sampling and although atmospheric turbulence obviously undergoes diurnal variations, over a short period of time the turbulence may be regarded as stationary. Obviously, at sunrise and sunset the source of heating to the atmosphere undergoes a sudden change and the turbulence will be non-stationary. At other times, provided there are no large-scale changes such as produced by an approaching cold front, the turbulence over a short period may be considered stationary.

In discussing the transport of atmospheric constituents, we consider the flux, or flow of the quantity through a unit cross-sectional area perpendicular to the flow, with units of, say, kilograms per second per square metre. When multiplied by the area concerned, this gives the total amount of the quantity (in this case, mass) transported per unit time through the area. Here we will consider the mass transport in the vertical direction through a horizontal area, A, as shown in Figure 2.27.

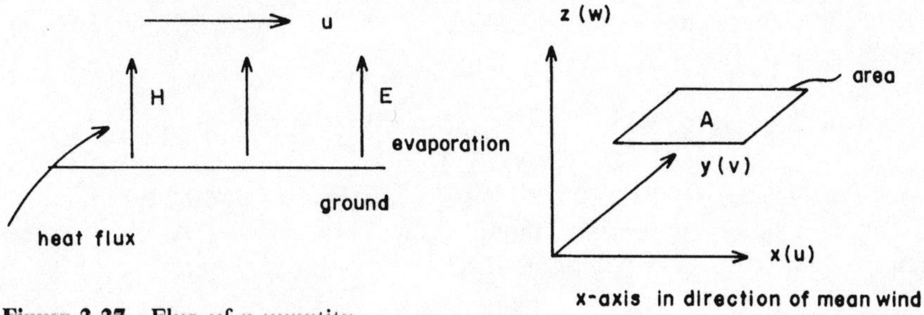

Figure 2.27 Flux of a quantity

The second part of Figure 2.27 shows the reference frame and we consider that the wind components are u, v, w, in the three coordinate directions. The area A lies at height z and has dimensions of $\Delta x \Delta y$. The volume of air passing through A per unit time travelling upwards is:

$$= A\ w.$$

Mass is volume times density so the mass of air passing through per unit time is

$$= \varrho A w$$

This is the upward flow of air in Kg s^{-1}.

Hence if we consider that there are q units of some entity per unit mass of air, we expect

$$= \varrho A wq$$

of that entity per second passing upward through area A. Thus the rate of transfer of the entity per unit area (or the flux of the entity) is

$$= \varrho wq.$$

This is the instantaneous, advective flux of the entity. Following the averaging principles given above, the mean flux can be represented as

$$= \overline{\varrho wq}.$$

Substituting approximately for ϱ, w and q, the Reynolds' rules of averaging simplify this average to:

$$= \overline{\varrho w}\ \bar{q} + \overline{(\varrho w)'\ q'}.$$

As there is no net vertical transport of air in the lower atmosphere over a sufficient time, we expect the average of ϱw to be zero to ensure conservation of mass. Therefore,

$$= \overline{(\varrho w)'\ q'} \approx \varrho\ \overline{w'q'},$$

since the density of the air, ϱ, is also approximately constant.

Taking q to be the specific humidity, the mass of water vapour per unit mass of dry air, the evaporation flux is

$$E = \varrho\ \overline{w'\ q'}.$$

Similarly the sensible heat flux is

$$H = C_p\ \varrho\ \overline{w'\ T'}$$

and the vertical flux of horizontal momentum is given by

$$\varrho \overline{w' u'}.$$

Note that ϱu is the horizontal momentum per unit volume and u, itself, is the horizontal momentum in the mean wind direction per unit mass.

The shearing stress, τ, is simply the rate of change of momentum per unit area or, through Newton's law, it is the force per unit area resulting from the vertical shear of the horizontal wind; therefore,

$$\tau = - \varrho \overline{w' u'}.$$

We have already shown that the shearing stress can be related to the vertical shear of the horizontal wind by the relationship

$$\tau = \varrho K_m \frac{\partial \overline{u}}{\partial z},$$

and by analogous arguments we expect that the heat flux H and the evaporation flux E are given by:

$$H = - \varrho K_H C_P \frac{\partial \overline{\theta}}{\partial z}$$

$$E = - \varrho K_w \frac{\partial \overline{q}}{\partial z},$$

where $\overline{u}$ is the mean wind speed, $\overline{\theta}$ the mean potential temperature and $\overline{q}$ the mean specific humidity.

The transfer coefficients defined in these equations for purely turbulent transport have direct molecular equivalents. However, although the molecular transport coefficients are not dependent on any microscopic length scales, the turbulent transfer coefficients (i.e. Ks) increase with height as the larger eddies become more effective a greater heights (see Figures 2.25 and 2.26). The fluxes are essentially constant with height when steady conditions occur and the gradients decrease with height.

The total rate of change undergone by any quantity A is equal to the local rate of change plus the advective rate of change. That is, given A (t, x, y, z) then

$$\frac{\mathrm{d}A}{\mathrm{d}t} \text{ (Lagrangian time derivative)}$$

$$= \frac{\partial A}{\partial t} + \frac{\partial A}{\partial x}\frac{\mathrm{d}x}{\mathrm{d}t} + \frac{\partial A}{\partial y}\frac{\mathrm{d}y}{\mathrm{d}t} + \frac{\partial A}{\partial z}\frac{\mathrm{d}z}{\mathrm{d}t}$$

$$= \frac{\partial A}{\partial t} + u\frac{\partial A}{\partial x} + v\frac{\partial A}{\partial y} + w\frac{\partial A}{\partial z}$$

Eulerian time derivative taken at fixed point in space + Eulerian space derivatives taken at fixed point in time (2.66)

The first total derivative (also called the substantial derivative) $\mathrm{d}A/\mathrm{d}t$, represents the total change one would observe travelling with a parcel moving at the velocity represented by the component speeds u, v and w. The term $\partial A/\partial t$ is the change that would be observed at a stationary point as time progressed, that is, it represents the local rate of change.

Consider an infinitesimal cube of side δx, δy, δz embedded in a flow field with velocity components, u, v, w at x, y, z as shown in Figure 2.28.

Mass flow into left hand x face per unit time $= \varrho u \, \delta y \, \delta z$.

Correspondingly, mass flow out of right hand x face per unit time $= \left(\varrho u + \frac{\partial(\varrho u)}{\partial x}\right) \delta x \, \delta y \, \delta z$,

allowing for the fluid speed, u, to vary with distance.

Hence the resulting increase of mass due to advection in the x direction

$$= -\frac{\partial(\varrho u)}{\partial x} \delta x \, \delta y \, \delta z.$$

Extending this to all three directions, the total increase in mass of the cuboid due to advection is

$$= -\left(\frac{\partial(\varrho u)}{\partial x} + \frac{\partial(\varrho v)}{\partial y} + \frac{\partial(\varrho w)}{\partial z}\right) \delta x \, \delta y \, \delta z.$$

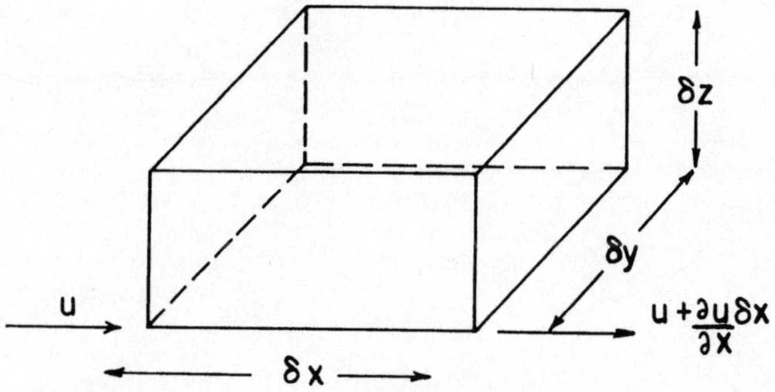

Figure 2.28 Conservation of mass

This increase in mass due to advection produces a mass accumulation rate in the cuboid that will lead to a change in density of the fluid in the cuboid.

This rate of mass increase is

$$= \frac{\partial \varrho}{\partial t} \delta x \; \delta y \; \delta z.$$

Hence, equating these two and dividing by the volume of the cuboid, $\delta x \; \delta y \; \delta z$, gives

$$\frac{\partial \varrho}{\partial t} = -\left(\frac{\partial(\varrho u)}{\partial x} + \frac{\partial(\varrho v)}{\partial y} + \frac{\partial(\varrho w)}{\partial z}\right),$$

where we have used some liberties to convert the equation to partial derivatives. Rearranging,

$$\frac{\partial \varrho}{\partial t} + \frac{\partial(\varrho u)}{\partial x} + \frac{\partial(\varrho v)}{\partial y} + \frac{\partial(\varrho w)}{\partial z} = 0.$$

That is,

$$\frac{\partial \varrho}{\partial t} + \varrho \frac{\partial u}{\partial x} + \varrho \frac{\partial v}{\partial y} + \varrho \frac{\partial w}{\partial z} + u \frac{\partial \varrho}{\partial x} + v \frac{\partial \varrho}{\partial y} + w \frac{\partial \varrho}{\partial z} = 0.$$

Substituting for the substantial derivative where $A = \varrho$ in (equation 2.66) leads to

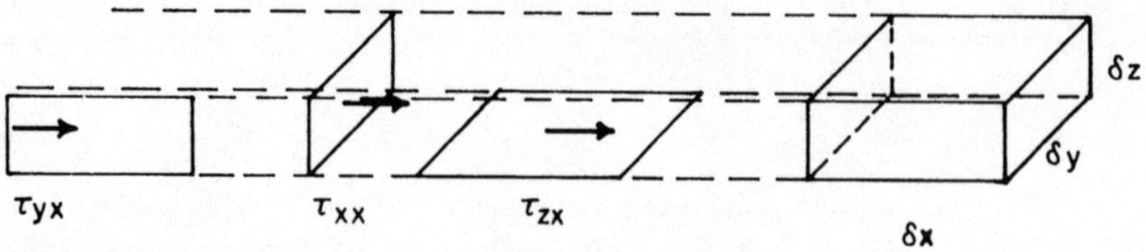

Figure 2.29 Shearing stress: forces acting on fluid element

$$\frac{d\varrho}{dt} + \varrho\left(\frac{\partial u}{\partial x} + \frac{\partial v}{\partial y} + \frac{\partial w}{\partial z}\right) = 0, \tag{2.67}$$

which is the equation of continuity.

The equations of motion are derived from a consideration of the force balances on a fluid element in the three component directions. In this balance we note that the net force in a given direction produces an acceleration, or a time rate of change of momentum, in accordance with Newton's law.

The forces acting on the outer faces of the fluid element are due to the shearing stresses, τ, which are forces per unit area that resist the motion. These have nine components since a force in a given direction (three possibilities) may act on a fluid element surface oriented in any direction, (three possibilities). We specify these stresses by subscripts, such that τ_{xy} designates the shear stress acting on an area element whose normal is in the x direction with a force directed in the y direction.

These τ values make up a tensor of rank 2. That is a matrix or table of values having 3^2 components. In the same scheme, a vector is a tensor of rank 1 or a line of values having 3^1 components (one for each of the x, y, z directions) and a scalar is a tensor of rank 0, having a single value.

Effectively these stresses are like pressures in which the forces may act in shear. They can be thought of as, for example, the force that opposes the sliding of playing cards during a shuffle. In this case it is simple sliding friction, but, indeed all materials attract and repel themselves and so tend to 'lock' together to some degree and resist motion. This happens in the air as well, though, of course, the restraining forces are small by ordinary standards. Yet these forces determine the wind profile near the ground, as we have seen.

Hence we consider only forces in the x direction and ignore the forces due to horizontal changes in pressure. The shearing force on the fluid element acting on face $\delta x\ \delta z$ at position y is given by

$$\tau_{yx}\ \delta x\ \delta z.$$

Similarly, the force acting on the $\delta x\ \delta z$ face at position $y + \delta y$ is given by

$$(\tau_{yx} + \frac{\partial\ \tau_{yx}}{\partial y}\ \partial y)\ \delta x\ \delta z.$$

This gives a net force acting in the x direction on both $\delta x\ \delta z$ faces of

$$-\ \frac{\partial\ \tau_{yx}}{\partial y}\ \delta y\ \delta x\ \delta z.$$

Similarly, if we consider the faces $\delta y\ \delta z$, the net force is

$$-\ \frac{\partial\ \tau_{xx}}{\partial x}\ \delta x\ \delta y\ \delta z,$$

which is, more or less, equivalent to a pressure force, though this stress term (as defined here) is not the pressure but the attractive or repulsive forces of nearby fluid elements.

The last net shear force in the x direction acts on the $\delta x\ \delta y$ faces and is the term

$$-\ \frac{\partial\ \tau_{zx}}{\partial z}\ \delta z\ \delta x\ \delta y.$$

Additional forces act on the element of fluid as body forces, that is forces per unit mass (such as gravity) that act throughout the body. Taking X as the x-component of the body force per unit mass, the net force in the x direction is

$$\varrho\ X\ \delta x\ \delta y\ \delta z\ -\ \frac{\partial\ \tau_{xx}}{\partial x}\ \delta z\ \delta x\ \delta y$$

$$-\ \frac{\partial\ \tau_{yx}}{\partial y}\ \delta y\ \delta x\ \delta z\ -\ \frac{\partial\ \tau_{zx}}{\partial z}\ \delta z\ \delta x\ \delta y.$$

However, from Newton, the net force in the x direction

$$= \mathrm{ma}$$

$$= \varrho\ \delta x\ \delta y\ \delta z\ \frac{\mathrm{d}u}{\mathrm{d}t},\ \text{since } \varrho \text{ is constant.}$$

Therefore,

$$\varrho\ \delta x\ \delta y\ \delta z\ \frac{du}{dt} = (\varrho X - \frac{\partial\ \tau_{xx}}{\partial x} - \frac{\partial\ \tau_{yx}}{\partial y} - \frac{\partial\ \tau_{zx}}{\partial z}\ \delta x\ \delta y\ \delta z$$

that is,

$$\frac{du}{dt} = \frac{-1}{\varrho}\left(\frac{\partial\ \tau_{xx}}{\partial x} + \frac{\partial\ \tau_{yx}}{\partial y} + \frac{\partial\ \tau_{zx}}{\partial z}\right) + X.$$

Similarly, considering the forces in the y and z directions,

$$\frac{dv}{dt} = \frac{-1}{\varrho}\left(\frac{\partial\ \tau_{xy}}{\partial x} + \frac{\partial\ \tau_{yy}}{\partial y} + \frac{\partial\ \tau_{zy}}{\partial z}\right) + Y,$$

$$\frac{dw}{dt} = \frac{-1}{\varrho}\left(\frac{\partial\ \tau_{xz}}{\partial x} + \frac{\partial\ \tau_{yz}}{\partial y} + \frac{\partial\ \tau_{zz}}{\partial z}\right) + Z.$$

These are the equations of motion, or simply Newton's law applied to a fluid, and are often called the Navier–Stokes equations.

The equation of continuity (2.67) showed us that

$$\frac{d\varrho}{dt} + \varrho\left(\frac{\partial u}{\partial x} + \frac{\partial v}{\partial y} + \frac{\partial w}{\partial z}\right) = 0,$$

which, for a fluid of constant density is:

$$\frac{\partial u}{\partial x} + \frac{\partial v}{\partial y} + \frac{\partial w}{\partial z} = 0,$$

and therefore multiplying through by u,

$$u\ \frac{\partial u}{\partial x} + u\ \frac{\partial v}{\partial y} + \frac{\partial w}{\partial z} = 0. \tag{2.69}$$

From the equation for the substantial derivative we know that

$$\frac{du}{dt} = \frac{\partial u}{\partial t} + u\ \frac{\partial u}{\partial x} + v\ \frac{\partial u}{\partial y} + w\ \frac{\partial u}{\partial z}, \tag{2.70}$$

where the last three terms on the right-hand side of the equation represent the inertial effects.

Adding equations 2.69 and 2.70 together leads to the resulting expression,

$$\frac{du}{dt} = \frac{\partial u}{\partial t} + \frac{\partial (u^2)}{\partial x} + \frac{\partial (uv)}{\partial y} + \frac{\partial (uw)}{\partial z},$$

which, from the Navier–Stokes equations is equal to,

$$\frac{du}{dt} = X - \frac{1}{\varrho}\left(\frac{\partial\,\tau_{xx}}{\partial x} + \frac{\partial\,\tau_{yx}}{\partial y} + \frac{\partial\,\tau_{zx}}{\partial z}\right).$$

But, if we substitute for the velocity as $u = U + u'$, where U is the mean velocity, take the mean and use the appropriate Reynolds' rules of averaging the above becomes:

$$\frac{\partial U}{\partial t} + \frac{\partial U^2}{\partial x} + \frac{\partial \overline{u'^2}}{\partial x} + \frac{\partial (UV)}{\partial y} + \frac{\partial (\overline{u'v'})}{\partial y} + \frac{\partial (UW)}{\partial z} + \frac{\partial (\overline{u'\ w'})}{\partial z}$$

$$= X - \frac{1}{\varrho}\left(\frac{\partial\,\tau_{xx}}{\partial x} + \frac{\partial\,\tau_{yx}}{\partial y} + \frac{\partial\,\tau_{zx}}{\partial z}\right)$$

from the Navier–Stokes equation in the x direction (equation 2.68). Further,

$$\frac{\partial U}{\partial t} + \frac{\partial U^2}{\partial x} + \frac{\partial (UV)}{\partial y} + \frac{\partial (UW)}{\partial z} = X - \frac{1}{\varrho}\{\frac{\partial}{\partial x}(\tau_{xx} - \varrho\overline{u'^2}) + \frac{\partial}{\partial y}(\tau_{yx} - \varrho\overline{u'\ v'}) + \frac{\partial}{\partial z}(\tau_{zx} - \varrho\overline{u'\ w'})\}. \quad (2.71)$$

This is now the same equation as Newton's law for a fluid with the mean velocity components U, V, W, but here we have additional stresses given by $\varrho\overline{u'^2}$, $\varrho\overline{u'\ v'}$, $\varrho\overline{u'\ w'}$ which result from the turbulence.

These additional stresses are known as the Reynolds' stress components $\varrho\overline{u'_i u'_j}$; when $i = j$ we have a normal component of the stress and when $i \neq j$ we have a shearing component. Thus the effect of the turbulence is to introduce additional stress terms into the equation of motion and so *the fluid experiences the turbulence through additional stress terms*.

These stress terms are often approximated with the eddy coefficient expression that relates the stress directly to the velocity shear. Alternatively, the stress terms may be measured directly with fast response instruments.

2.9 Statistical measures

Fluctuations, such as u', may be considered as an aggregate of sinusoidal waves, each with a frequency (n cycles/second), a period ($2\pi/n$ seconds), a wavelength (λ m) and a wave number (K rad m^{-1}). The dimensions of wave number are frequently abbreviated to m^{-1} and it is important not to confuse cycles m^{-1}, also abbreviated as m^{-1}, with rad m^{-1} (2π rad $=$ 1 cycle).

The frozen turbulence hypothesis of G. I. Taylor (1938) assumes that the pattern of turbulence remains unchanged as the stream passes the fixed instrument. This implies that the time scale, t, can be interpreted as the space scale, ut, in the direction of flow and asserts that the wavelength is the product of the period and the mean wind.

$$\text{(wavelength)} \qquad \lambda = \frac{2\pi\bar{u}}{n}$$

$$\text{(wave number)} \quad K = \frac{2\pi}{\lambda} = \frac{n}{\bar{u}}$$

It is not implied that all eddies move along with the mean flow velocity and experience no distortion. Individual eddies are in a constant state of growth and decay, but, for low levels of turbulence, the above equations represent a useful approximation (Munn, 1966).

The time series of the turbulence is a sequence of values as a function of time or other independent variable and hence we may write:

$$u' = u'(t).$$

One statistical property of this series is the correlation coefficient

$$R_{uv} = \frac{\overline{u'\,v'}}{(\overline{u'^2}\;\overline{v'^2})^{1/2}},$$

which relates the two horizontal wind components, where the means are taken over the entire data series. Its properties are such that $R_{uv} = 1$ for perfect correlation, $R_{uv} = -1$ for perfect inverse correlation, and $R_{uv} = 0$ for no relationship. Similarly, the auto-correlation coefficient may be defined as

$$R_u\,(\sigma) = \frac{\overline{u'(t)\;u'(t+\sigma)}}{\overline{u'^2}},$$

where σ is not the standard deviation, but is the set time lag between equivalent measurements of u at a given point in space. As such the auto-correlation gives a measure as to how well observations separated by a set time internal are correlated.

In a similar manner, we may define a cross-correlation coefficient,

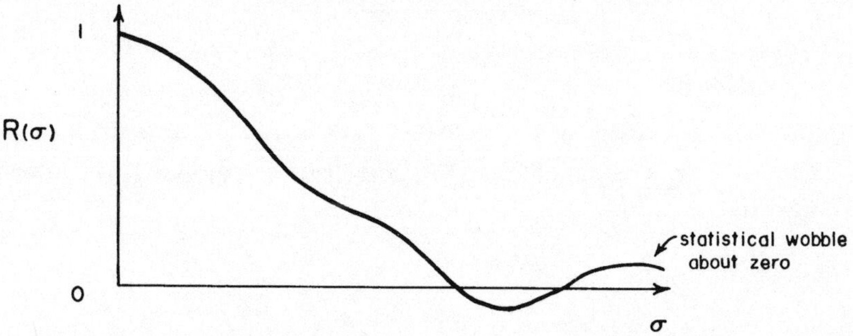

Figure 2.30 Typical form of the auto-correlation function

$$R_{uv} = \frac{\overline{u'(t)\ v'(t + \sigma)}}{(\overline{u^2})^{1/2}\ (\overline{v'^2})^{1/2}}$$

or

$$R_{uw} = \frac{\overline{u'(t)\ w'(t + \sigma)}}{(\overline{u'^2})^{1/2}\ (\overline{w'^2})^{1/2}}.$$

In these equations u', v' and w' are strictly defined with the mean $\overline{u}$ evaluated over the relevant time interval $0 < t < T - \sigma$, where T is the total time length of the record.

The general form of the auto-correlation function, also known as the correlogram, is shown in Figure 2.30. Note the perfect correlation at $\sigma = 0$ when, of course, both $u'(t)$ and $u'(t + \sigma)$ are identical and so must show a perfect correlation. For larger time lags the function becomes smaller and approaches zero. If, indeed, one can think of a 'mean eddy' this zero crossing (i.e. $\sigma = \sigma_0$) would represent the 'centre' of the eddy. The radius would be approximated by $\sigma_0/\overline{u}$ if the turbulence were reasonably steady in passing the point of measurement (Taylor frozen turbulence concept). Further, since we expect $R_u(\sigma)$ to be the same as $R_u(-\sigma)$ the function is even and has a zero slope at the origin. This also means that the function can be approximated by a series of even functions and, indeed, it is often separated into a Fourier cosine series. As σ increases beyond σ_0 we expect R_u to become negative and, indeed the function exhibits a statistical wobble about zero. This is most apparent when there are few measurements at large σ, as there always must be because of a finite record length. In the limit of large σ, any correlation must only relate to a single eddy of large size and cannot have any statistical validity.

The crossing point σ_0 is generally ill-defined, so a greater appreciation of the scale of turbulence can be achieved by using the integral of the curve as the time scale. We define t_a, the integral scale as,

$$t_a = \int_0^\infty R(t)\, dt$$

For example, if it is assumed that the correlogram can be represented by an exponential function,

$$R(t) = e^{-ct}$$

then

$$t_a = \int_0^\infty e^{-ct}\, dt$$

$$= \frac{1}{c}.$$

In this instance, the integral scale is the same as the time at which R drops to 1/e (= 0.368).

Realising that the correlation coefficient can be represented by an even function and that the function is non-repetitive, it is usual to decompose it into a Fourier integral,

$$R(t) = \int_0^\infty F(n) \cos 2\pi nt\, dn,$$

which is effectively an infinite series expression for R in terms of cosines. The $F(n)$ are the Fourier coefficients of the terms with n, the frequency of a given cosine form. The orthogonal properties of the Fourier series allow the computation of the $F(n)$. They are given by

$$F(n) = 4 \int_0^\infty R(t) \cos 2\ nt\, dt,$$

where $F(n)$ is a continuous function, a Fourier transform of $R(t)$. In fact, the similarity of these equations leads to the concept of $R(t)$ and $F(n)$ being 'Fourier transform pairs'. In fact, the transformation converts the time variable t into the frequency variable n.

The product $u'(t)\, u'(t + \sigma)$ that makes up the correlation coefficient is a measure of the kinetic energy (K.E. = ½ mv^2), since it involves a velocity squared. Indeed the correlogram partitions the energy according to

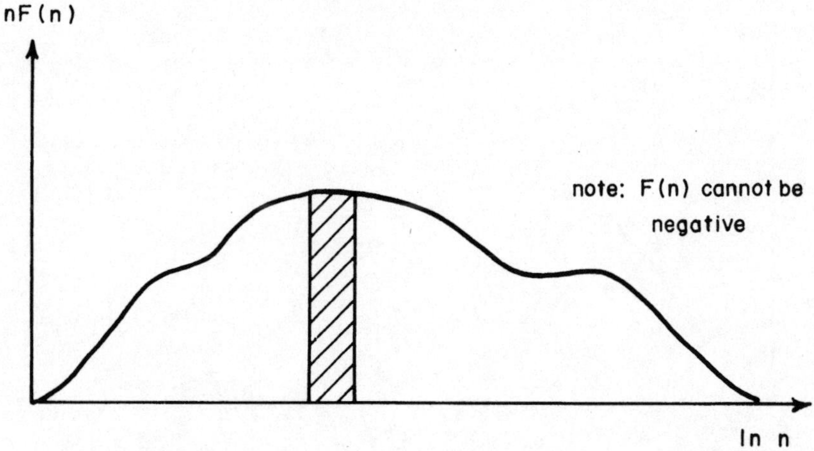

Figure 2.31 Typical form of the spectrum function

the 'time lag' σ between observations. The Fourier transform, $F(n)$, correspondingly partitions the energy between frequencies. Hence the spectrum, shown in Figure 2.31, illustrates how this energy is distributed among the various turbulent frequencies. A few very large fluctuations may yield the same value of $\overline{u'^2}$, as would many small fluctuations and hence the spectrum function and spectrum analysis may identify the importance of the various contributing frequencies. We have already noted that under stable conditions, atmospheric spectra illustrate little variance whereas instability leads to a shift of the spectra to lower frequencies.

The shaded area in Figure 2.31 represents the contribution of that frequency band to the total spectra. Hence by plotting the spectra it is possible to illustrate the relevant importance of various frequency bands to the total spectra.

The normal representation of spectra on a logarithmic scale preserves equality of area in corresponding wave bands. Kolmogorov has shown that under local isotropy (that is high frequency turbulence is isotropic although the low frequency fluctuations may not be), we can calculate the energy dissipation rate defined as the change of turbulent kinetic energy due to viscosity effects:

$$\epsilon = 15\,\nu \int_0^\infty K^2\,F(K)\,dK,$$

where ϵ is the energy dissipation rate, ν is the kinematic viscosity of the

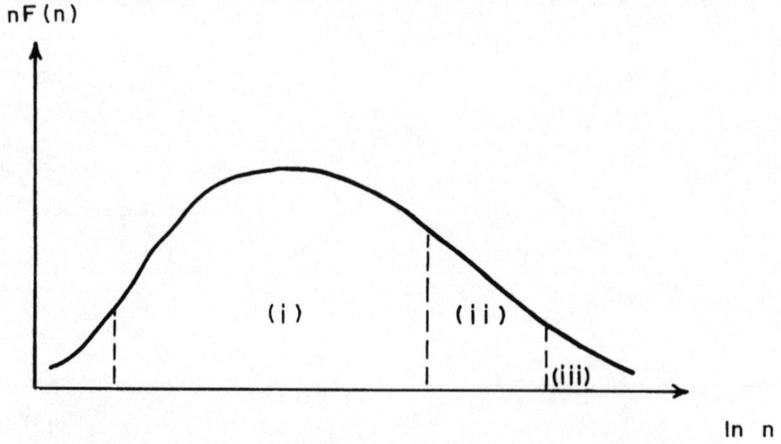

Figure 2.32 Typical shape of the spectra of atmospheric turbulence

fluid, and K is the wave number. Thus a plot of $K^2\ F(K)$ against K, or of $K^3\ F(K)$ against ln K, will show the contribution of different wave numbers to the energy dissipation rate.

Turbulent flow is dissipative and therefore needs a continuous supply of energy. In fact, turbulent flow may be thought of as a regime of eddies wherein there is a cascade of energy from the largest to the small eddies, where the energy is dissipated by molecular interactions in the form of heat. That is, the energy enters the spectrum at relatively low frequencies (large eddies), and is transferred to higher and higher frequencies (small eddies), until it is finally dissipated.

Thus the general shape of the spectra of atmospheric turbulence may be represented as shown in Figure 2.32, where

(i) is the region of input of turbulent energy and the energy containing eddies;
(ii) is the conservative range as energy which is in the turbulence is passed down from the larger to smaller eddies;
(iii) is the viscous range where turbulent energy is dissipated by viscosity.

Since a wind sensor does not respond completely to high frequency fluctuations, experimental records always include unavoidable smoothing. If the smoothing time is S, it may be shown (i.e. Pasquill, 1974; Munn, 1966) that the variance of u diminishes according to the relationship,

$$\overline{u'^2_S} = \overline{u'^2} \int_0^\infty \frac{F(n) \sin^2 \pi nS}{(\pi nS)^2} \, dn$$

where $\overline{u'^2_s}$ is the experimental realisation of $\overline{u'^2}$. For increasing S, $\overline{u'^2_s}$ approaches zero as more and more of the fluctuations are smoothed away. Decreasing S, on the other hand, leads to $\overline{u'^2_s}$ approaching $\overline{u'^2}$ since the effect of the smoothing is reduced.

Since the wind record under consideration is not of infinite length, some of the low frequency eddy energy will also be lost. If the length of the record is T (minutes), it may be shown (Munn, 1966) that

$$\overline{u'^2_T} = \overline{u'^2} \int_0^\infty F(n) \left(1 - \frac{\sin^2 \pi n T}{(\pi n T)^2}\right) \mathrm{d}n$$

Hence the joint effect of smoothing time, S, and discrete length of record, T, is given by,

$$\overline{u'^2_{S,T}} = \overline{u'^2} \int_0^\infty F(n) \left(\frac{\sin^2 \pi n S}{(\pi n S)^2} - \frac{\sin^2 \pi n T}{(\pi n T)^2}\right) \mathrm{d}n. \tag{2.72}$$

Theoretically, this implies that the real variance of standard deviation of a turbulent eddy is never completely observed. In practice, a close approach to the true value may be achieved if the sampling duration and averaging times are, respectively, long enough and short enough in comparison with the characteristic time scale of the spectrum. Except for the vertical component very near the ground when the fluctuations are concentrated at relatively high frequency, the second of the above requirements is not difficult to achieve. Adequate sampling duration is, however, difficult to realise within the period over which conditions remain constant, especially for the horizontal components and for the vertical component at reasonable heights, since in both cases the time scales are relatively large.

Many of the fundamental theories of fluid mechanics are cast in the Lagrangian reference frame, whereas the vast majority of meteorological observations are carried out in the Eulerian system. The Lagrangian reference frame moves with the motion whereas the Eulerian reference frame remains stationary. Hence, there is a need to relate the two systems, and, in particular, to find relationships between the parameters in the different frames of reference. Quantities such as

$$u', \overline{u'^2}, \overline{w'^2} \text{ and } \overline{u'w'}$$

are the same in either system although the spectra and correlation functions are different. However, it is generally assumed that

$$R_L(\xi) = R_E(t),$$

where $R_L(\xi)$ and $R_E(t)$ are the Lagrangian and Eulerian correlation

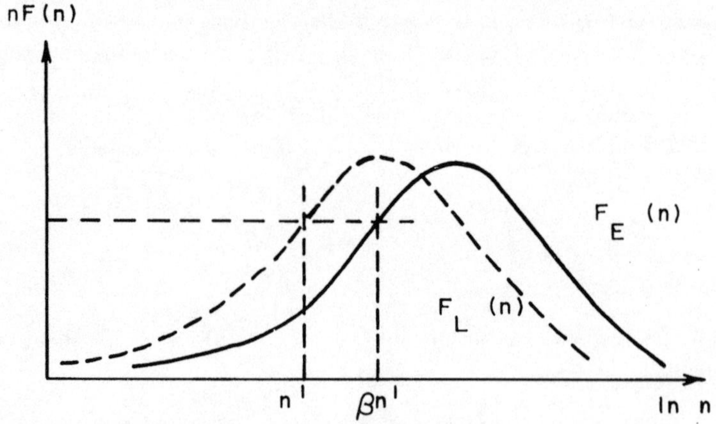

Figure 2.33 Lagrangian and Eulerian reference frames — spectral function

functions and ξ and t are the two time scales, with $\xi = \beta t$ (Pasquill, 1974). Since the spectral function is the Fourier transform of the correlation function it follows that

$$F_L(n) = \beta F_E(n).$$

Figure 2.33 shows the relationship between the two frames of reference for the spectra (Pasquill, 1974). It can be seen that the Lagrangian spectral function is simply shifted along the abscissa with respect to the Eulerian spectral function.

Experimental data support the above as useful assumptions although there is some question regarding the value of β. For example, Hanna (1981) suggests that it is inversely proportional to turbulence intensity and this implies an increase of β as stability increases.

2.10 Boundary-layer scaling

The top of the atmospheric boundary layer can be defined as the lowest level in the atmosphere at which the ground surface no longer influences the meteorological variables through the turbulent transfer of mass (Pielke, 1984). During the day it roughly corresponds to the height to which pollutants are mixed (the mixing height), whereas at night it is typically an order of magnitude smaller than the daytime value over land. Over the sea, the diurnal variation is much less as the sea surface temperature does not experience the extreme diurnal variations of the land surface.

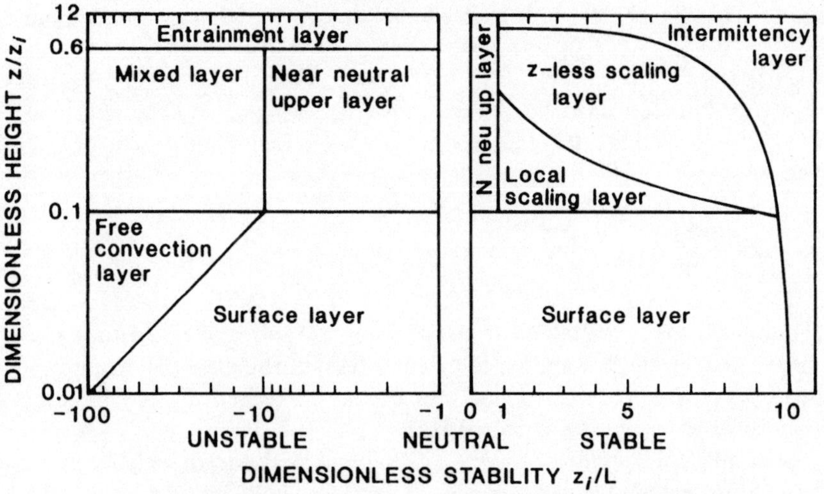

Figure 2.34 Scaling regimes within the atmospheric boundary layer (after Holtslag and Nieuwstadt, 1986 and reproduced in Gryning *et al.*, 1987)

The turbulent structure of the boundary layer has been characterised in terms of three length scales: z the height above the surface, z_i the mixing height and L the Monin–Obukhov length scale (Holtslag and Nieuwstadt, 1986). As we have already seen L is a characteristic of the flow whereas z limits the physical size of the eddies and z_i specifies the vertical limit of their propagation. From the dimensional analysis, these can be expressed as two independent dimensionless ratios, a relative height z/z_i and a measure of the boundary-layer stability z_i/L.

Holtslag and Nieuwstadt (1986) used these to define the scaling regimes shown in Figure 2.34 and identified the parameters that determine the structure of turbulence. For unstable conditions ($L < 0$) they defined the surface layer, the free convection layer, the mixed layer, the near neutral upper layer and the entrainment layer, whereas for stable conditions ($L > 0$) the layers were defined as the surface layer, the local scaling layer, the z-less scaling layer, the near neutral upper layer and the intermittency layer. Little is known of the near-neutral upper layer, entrainment layer or the intermittency layer and consequently the dividing lines between the various regimes are somewhat arbitrary. Nevertheless, they provide a convenient framework for scaling the turbulent characteristics of the atmosphere.

In the surface layer, wind shear and the radiational heating or cooling of the surface are the dominant controls over turbulence. Turbulence within this layer is scaled by Monin–Obukhov similarity theory and, as we have already seen, the controlling parameters are H_0, u_*, g/T and z.

In the mixed layer, intense vertical mixing tends to leave conserved variables such as potential temperature and humidity nearly constant with height. This layer is characterised by strong downdrafts and updrafts leading Deardorff (1970) to suggest that the parameters g/T_0, z, z_i and $\overline{w\theta_0}$ alone determine the properties of turbulence. That is within the mixed layer turbulence is determined by the heat flux, $(g/T)\overline{w\theta_0}$, and the depth of the mixed layer, z_i. This led him to define a characteristic velocity scale w_* given as

$$w_* = \left(\frac{g\overline{w\theta_0}z_i}{T}\right)^{1/3}$$

The mixed layer scaling hypothesis, using w_* and z_i, has led to a marked improvement in understanding the structure of turbulence (Wyngaard, 1988) and dispersion within this region can be parameterised directly in terms of these variables (Gryning *et al.*, 1987).

Under neutral conditions, $L \rightarrow \infty$ implying that the turbulence is scaled by the distance above the ground, z, which is the only available length scale, since the mixing height has no meaning under neutral conditions. Thus the size of the turbulent eddies is determined solely by their height above the surface. If the stability increases, the turbulent eddies will be suppressed and become much smaller than their neutral counterparts. As they become smaller, the size of the eddies, or the degree of suppression of turbulence, is determined by the increase in stability and thus the eddy sizes become independent of the height above the ground and are related directly to the stability. This is the z-less scaling layer. In the local scaling layer, turbulence is also scaled by the local values of the momentum and heat fluxes (Nieuwstadt, 1984) but the proximity of the surface also implies a dependence on z.

In the convective surface layer, vertical σ_w and horizontal, σ_u, σ_v turbulence components have been represented as

$$\sigma_w/u_* = 1.3[1 + 3(-z/L)]^{1/3}$$

$$\sigma_u/u_* = \sigma_v/u_* = [12 + 0.5(-z_i/L)]^{1/3}$$

and these expressions form a convenient interpolation between the neutral and strongly convective limits. Under neutral conditions $\sigma_w \propto u_*$ whereas at the free convection limit, $\sigma_w = 1.34(g\overline{w\theta_0}z/\theta_0)^{1/3}$, whereas σ_u, $\sigma_v \propto u_*$ under neutral conditions and $\propto w_*$ under strong convection. Thus with strong convection in the surface layer, the horizontal turbulence components are not a function of stability (z/L) but are dependent on the velocity scale, w_*, of the large eddies (Weil, 1988c) and hence do not obey surface-layer scaling.

Chapter 3

ATMOSPHERIC DIFFUSION

Small particles or droplets released into the atmosphere become dispersed or separated from one another under the influence of turbulence. This effect has been noted several times and is, more or less, analogous to molecular diffusion by the random motions of molecules. In turbulence, however, random eddy motions play this role. Hence the phenomenon is known as turbulent diffusion and, indeed, to a first approximation it tends to be 'down-the-gradient'. However, turbulent diffusion in the atmosphere has not yet been uniquely formulated in the sense that a single basic physical model capable of explaining all the significant aspects of the problem has not been proposed. Instead, there are two alternative approaches which each have areas of utility with little overlap; they are the gradient transport approach and the statistical approach. Diffusion at a fixed point in the atmosphere, according to gradient transport, is proportional to the local concentration gradient. Consequently, it could be said that this theory is Eulerian in nature in that it considers properties of the fluid motion relative to a spatially fixed coordinate system. On the other hand, the statistical approach considers motions following individual fluid particles and thus can be described as Lagrangian.

3.1 Turbulent gradient transport

In the gradient-transfer approach it is assumed that turbulence causes a net movement of material down the gradient of material concentration, at a rate which is proportional to the magnitude of the gradient. The proportionality factor is of course analogous to the coefficients of viscosity or conductivity in the familiar laws for the molecular transfer of momentum or heat in laminar flow. Generally,

$$F_S = -A\frac{\partial\bar{S}}{\partial n}$$

$$= -\varrho K\frac{\partial\bar{S}}{\partial n}, \qquad (3.1)$$

where F_s is the eddy flux, that is, the rate of eddy transfer per unit area across a fixed surface perpendicular to the flow and $\partial\bar{S}/\partial n$ is the gradient of the property, with $\bar{S}$ being the mean quantity per unit mass of air.

The negative sign in these equations is consistent with down-gradient flux and this definition in terms of an exchange coefficient A or an eddy diffusivity K has already been introduced for the case of momentum. Turbulent transfer following such a law is referred to as a simple turbulent diffusion process.

The instantaneous value of the quantity may be represented in terms of a mean and an eddy fluctuation from the mean so that the overall vertical transport flux through a horizontal plane is

$$\varrho\overline{wS} = \varrho\bar{w}\bar{s} + \varrho\overline{w'S'}, \qquad (3.2)$$

where upward velocities and transports are conventionally taken as positive. The first term arising from any mean vertical motion is customarily taken to be zero at low heights over level uniform terrain, as there is no net vertical motion in the atmosphere. Hence, to a good approximation, the total flux is equivalent to the second term, the eddy flux.

The differential equation, which has been the starting point of most mathematical treatments of diffusion from sources, is a generalisation of the classical equation for conduction of heat in a solid and is essentially a statement of the conservation of the suspended material (Slade, 1968). Denoting the local concentration by q units of mass per unit volume of fluid and assuming the fluid to be incompressible, the equation of continuity can be written

$$\frac{\partial q}{\partial t} = -\left(\frac{\partial(uq)}{\partial x} + \frac{\partial(vq)}{\partial y} + \frac{\partial(wq)}{\partial z}\right). \qquad (3.3)$$

The quantities u, v, w and q are represented as the sum of a mean and eddy fluctuation and are substituted into equation 3.3. The terms are expanded, the equation is averaged; we find that:

$$\frac{\partial \bar{q}}{\partial t} + \bar{u}\frac{\partial \bar{q}}{\partial x} + \bar{v}\frac{\partial \bar{q}}{\partial y} + \bar{w}\frac{\partial \bar{q}}{\partial z}$$

$$= -\left(\frac{\partial(\overline{u' q'})}{\partial x} + \frac{\partial(\overline{v' q'})}{\partial y} + \frac{\partial(\overline{w' q'})}{\partial z}\right). \tag{3.4}$$

The eddy flux terms may now be replaced by the simplest gradient transfer forms and equation 3.4 becomes

$$\frac{d\bar{q}}{dt} = \frac{\partial}{\partial x}(K_x \frac{\partial q}{\partial x}) + \frac{\partial}{\partial y}(K_y \frac{\partial q}{\partial y}) + \frac{\partial}{\partial z}(K_z \frac{\partial q}{\partial z}), \tag{3.5}$$

where equation 2.66 has been used to replace the left-hand side of equation 3.4.

This equation allows for differences in the eddy diffusivities in the component directions, that is for anisotropic diffusion, and also for spatial variations of these diffusivities. If the *K*s are constant, independent of *x*, *y* or *z*, the diffusion follows Fick's law of diffusion and is called Fickian.

Thus for the one dimensional case, Fick's law of turbulent diffusion is

$$\frac{d\bar{q}}{dt} = K\frac{\partial^2 \bar{q}}{\partial x^2}, \tag{3.6}$$

but if the medium is stationary, this equation reduces further to,

$$\frac{\partial \bar{q}}{\partial t} = K\frac{\partial^2 \bar{q}}{\partial x^2}, \tag{3.7}$$

which is the time-dependent one-dimensional diffusion equation without advection.

The case with transport, equation 3.6 can be rewritten in the form,

$$\frac{\partial \bar{q}}{\partial t} + \bar{u}\frac{\partial \bar{q}}{\partial x} = \mathrm{K}\frac{\partial^2 \bar{q}}{\partial x^2},$$

where there is transport in only one dimension with $\bar{v} = \bar{w} = 0$. The first term on the left-hand side of this equation represents the change of q with time at a selected, fixed position in space. The second term represents the q advected at velocity $\bar{u}$.

If $\bar{u}$ is zero, this equation is the same as equation 3.7. However, the same equation results if we move the coordinate system along with the flow at velocity $\bar{u}$. We consider a puff or point source of pollutant and measure our concentrations relative to the centre of the puff. The resulting solution is identical to the stationary case.

In order to solve equation 3.6, it is necessary to specify the boundary conditions and for a point source of pollutant these are:

(i) the concentration at all points goes to zero as time after the release of pollutant approaches infinity,

$$\bar{q} \to 0 \text{ as } t \to \infty \qquad (-\infty < x < +\infty);$$

(ii) the concentration is zero as time after the release approaches zero for all points except the source,

$$\bar{q} \to 0 \text{ as } t \to 0 \qquad (\text{for all } x \text{ except } x = 0);$$

(iii) the total mass of pollutant present is equal to the amount released,

$$\int_{-\infty}^{\infty} \bar{q} \, dx = Q,$$

where Q is the source strength, that is the total release of $\bar{q}$ from the source located at $x = 0$.

The solution to this problem is a Gaussian function of the general form

$$\bar{q} = \frac{1}{at^{1/2}} \exp\left(-\frac{bx^2}{t}\right),$$

where the constants a and b can be evaluated from the set boundary conditions. The solution to equation 3.6 for an instantaneous point source of strength Q is

$$\frac{\bar{q}}{Q} = \frac{1}{(4\,Kt)^{1/2}} \exp\left(\frac{-x^2}{4\,Kt}\right). \tag{3.8}$$

This solution applies to an atmosphere in which $\bar{u}$ is constant, $v = w = 0$, and for which the coordinates are thought of as moving with the mean wind $\bar{u}$.

Equation 3.8 may be extended to three dimensions to give

$$\frac{\bar{q}\,(r,t)}{Q} = \frac{1}{(4\pi Kt)^{3/2}} \exp\left(\frac{-r^2}{4\,Kt}\right), \tag{3.9}$$

where

$$K = K_x = K_y = K_z$$

and

$$r^2 = x^2 + y^2 + z^2.$$

Further generalisations for the case of non-isotropic diffusion (i.e. $K_x \neq K_y \neq K_z$) gives

$$\frac{\bar{q}\,(x,y,z,t)}{Q} = \frac{1}{(4\pi t)^{3/2}\,(K_x K_y K_z)^{1/2}} \exp\left(\frac{-1}{4t}\left(\frac{x^2}{K_x} + \frac{y^2}{K_y} + \frac{z^2}{K_z}\right)\right).$$

These equations represent the fundamental building blocks of Fickian theory. Integration of the instantaneous point source solution with respect to space yields equations for instantaneous volume sources (such as bomb bursts), whereas integration with respect to time gives the continuous point source solution (Slade, 1968; Pasquill, 1974; Hanna *et al.*, 1982).

The assumption of constant-eddy diffusivity, although it may be of considerable use in the free atmosphere, can hardly apply to the planetary boundary layer (see Figure 2.25), characterised as it is by pronounced shear of the mean wind and large variations in the vertical temperature gradients due to the heat flux.

Equation 3.5 may be simplified if we assume steady state conditions, that is $\partial\bar{q}/\partial t = 0$, take an infinite crosswind line source for which, effectively,

$$\frac{\partial}{\partial y}\left(K_y \frac{\partial \bar{q}}{\partial y}\right) = 0,$$

with the mean wind blowing along the x axis so that $\bar{v} = \bar{w} = 0$, and assume that x transport by the mean flow greatly outweighs the eddy flux in that direction. Under these conditions equation 3.5 reduces to

$$\bar{u}\,\frac{\partial \bar{q}}{\partial x} = \frac{\partial}{\partial z}\left(K_z \frac{\partial q}{\partial z}\right). \tag{3.10}$$

For the lower atmosphere, in adiabatic conditions, we have seen that the wind velocity varies with the logarithm of the height, but such a variation proves mathematically intractable in the manipulation of the above equation. A more manageable form has been obtained by adopting a power law form of the wind profile in which it is assumed that

$$K_z(z) = K_i(z/z_i)^n$$

$$\bar{u}(z) = \bar{u}_i(z/z_i)^m$$

where $\bar{u}_i$ and K_i are values of $\bar{u}$ and K_z at a fixed reference height z_i. With these forms equation 3.10 is soluble. In particular, for a constant wind with height ($m = 0$) and K varying linearly with height ($n = 1$) and in fact, given explicitly by

$$K = ku_{*}z,$$

the solution to equation 3.10 is

$$q = \frac{Q}{ku_{*}x} \exp\left(\frac{-uz}{ku_{*}x}\right).$$

3.2 Statistical theories of turbulent diffusion

The statistical approach studies the histories of the motion of individual fluid parcels and tries to determine from these the statistical properties necessary to represent diffusion, rather than considering the material or momentum flux at a fixed space point. Perhaps the simplest model of diffusion is the well-known random walk or the drunkard's walk, which is essentially a simple discrete step, stochastic diffusion model. Each step in the drunkard's walk is independent of his previous motion, and in the analogous diffusion model the movement of a pollutant is uncorrelated with its previous motion. The molecular analogue of this model describes Brownian diffusion and for large time-intervals the distribution approaches the normal distribution or Gaussian distribution. Unlike Brownian motion, however, actual turbulent atmospheric motions tend to be rather highly self-correlated as we have already seen.

The uncorrelated kind of diffusion process described by the Gaussian distribution corresponds closely to Fickian diffusion; consequently it must be governed by a parabolic type of differential equation, such as

$$\frac{\partial \bar{q}}{\partial t} = K \frac{\partial q^2}{\partial x^2}.$$

Solutions of parabolic equations have the mathematical character that some effect is felt everywhere except at the initial instant, $t = 0$. Hence the implication of the direct solution to the above equation is that diffusion proceeds in some sense with infinite velocity. Generalisations to a more realistic discrete-step diffusion model in which successive events are correlated (random walk with memory) have been carried out by numerous workers and these indicate that atmospheric diffusion should obey the 'telegrapher's equation'. Since the telegrapher's equation is hyperbolic rather than parabolic, it describes diffusion that proceeds at a finite

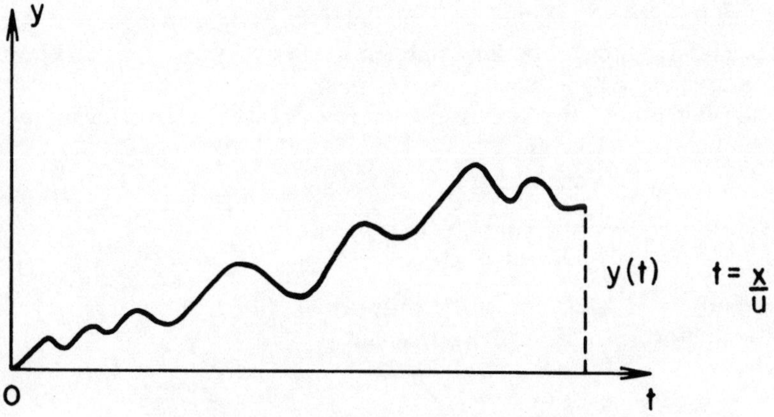

Figure 3.1 Displacement of individual particle, undergoing turbulent motion, as a function of time

velocity. Thus there will be a definite limit to the distance that fluid particles can disperse in a given amount of time; although this model is more realistic the practical difference is not great (Slade, 1968).

In stationary, homogeneous atmospheric flow, the distance, y, that an individual particle is carried away from an origin by turbulent wind fluctuations, v', during a time interval, t, is equal to

$$y(t) = \int_0^t v'(t_i)\,\mathrm{d}t_i \tag{3.11}$$

as shown in Figure 3.1. The simplest meaningful statistical measure of this irregular, random process is the mean-square diffusion that would result from a large number of repetitions. That is, the variance of the resulting distribution of particles along the y axis. Squaring both sides of equation 3.11 and taking the average over many repetitions of the experiment (that is the statistical or ensemble average) yields

$$\overline{y^2}(t) = 2\,\overline{v'^2} \int_0^t \int_0^t R(\xi)\,\mathrm{d}\xi\,\mathrm{d}t_i, \tag{3.12}$$

that is

$$\sigma_y^2 = 2\sigma_v^2 \int_0^t \int_0^t R(\xi)\,\mathrm{d}\xi\,\mathrm{d}t,$$

which is the classical result of Taylor, although it has since been generalised to three dimensions by Batchelor (Pasquill and Smith, 1983; Weil, 1988c). The function $R(\xi)$ is called the one-point Lagrangian velocity correlation coefficient; Lagrangian because it refers to the velocity of a particle rather than the velocity at a fixed point in space.

Following our earlier discussion we can define the Lagrangian velocity correlation coefficient as

$$R(\xi) = \frac{\overline{v'(t)\, v'(t + \xi)}}{\overline{v'^2}}.$$

Since $R(0) = 1$ and, for sufficiently small diffusion time $R(t) \approx 1$, it follows that when t is sufficiently small,

$$\sigma_y^2 \approx \sigma_v^2\, t^2. \tag{3.13}$$

When t is large, it may be supposed that the auto-correlation function R must approach zero sufficiently rapidly that

$$\sigma_v^2 \lim_{t \to \infty} \int_0^t R(t_i)\, dt_i = K_i,$$

where K_i is some constant. That is, the particle must ultimately 'forget' its original motion. Consequently, for large diffusion times,

$$\sigma_y^2 \approx 2\, K_i\, t. \tag{3.14}$$

The quantity

$$\int_0^\infty R\, dt$$

defines a time-scale characteristic of the turbulence known as the Lagrangian integral time scale,

$$T_L = \int_0^\infty R(t)\, dt,$$

which is a characteristic time for the turbulent diffusion. The integral time scale characterizes the eddies carrying the bulk of the turbulent kinetic energy and performing most of the turbulent transport. An alternative derivation has been given by Venkatram (1988a), which serves to highlight its role in our understanding of the diffusion process.

It is apparent from equation 3.14 that K_i has the dimensions of a diffusivity and therefore it might be expected that K_i plays a part similar to that of the K of Fickian theory. In particular, it appears reasonable that the Fickian theory, in which K is constant, should apply when the diffusion time, t, is large compared to the Lagrangian integral time scale. Thus the Fickian theory should only be applied to large diffusion times and is not appropriate for small diffusion times when equation 3.13 describes the process.

Unfortunately, there appears to be no basic way of evaluating the precise points at which the limits of Taylor's theorem for small and large times will apply in the atmosphere. In fact, the most reliable knowledge of the form of $R(\xi)$ has been inferred from diffusion experiments and its form is not known with precision. For example, one commonly used form for σ_y that satisfies the two limits is

$$\sigma_y = \sigma_v t/(1 + t/2T_L)^{0.5}.$$

In applying this equation, T_L is treated as an empirical parameter that can be derived by fitting the equation to tracer observations (Draxler, 1976; Gifford, 1987).

Taylor's theorem can also be written in terms of the eddy energy spectrum as

$$\sigma_y^2 = \sigma_v^2\, t^2 \int_0^\infty F(n)\, \frac{\sin^2(\pi n t)}{(\pi n t)^2}\, dn, \qquad (3.15)$$

where n is the frequency and $F(n)$ is the Lagrangian eddy-energy spectrum, the Fourier transform of the Lagrangian velocity correlation function $R(\xi)$.

It can be seen from this that σ_y^2 depends on the entire energy spectrum for any values of t and also as t increases the diffusion becomes dominated by the lower frequency contributions to $F(n)$. This follows because the spectrum in equation 3.15 is weighted by the function

$$W(n,t) = \frac{\sin^2(\pi n t)}{(\pi n t)^2},$$

which is largest for small values of nt and rapidly approaches zero for other values. Hence it appears that the large eddies, that is the Fourier components of the motion having a low frequency, dominate atmospheric diffusion. For continuous-source diffusion this means that the diffusion is only influenced by eddies with diameters roughly equal to and larger than σ. In other words as the plume grows the influence of small eddies will decrease, since oscillations of particles deep within the plume do not contribute to the diffusion of the plume as a whole.

The integrand of equation 3.15 is similar in form to the expression by which a computed turbulence energy spectrum is corrected for the effect of averaging the raw data over a time interval a, as shown in Chapter 2. That is,

$$F_a(n) = F(n)\left(\frac{\sin \pi na}{\pi na}\right)^2,$$

where $F_a(n)$ is the instrumental observed spectrum. Hence if the averaging interval is selected to be equal to the time of travel, or diffusion time, t, it follows that

$$\sigma_y^2 = \overline{v'^2}\, t^2 \int_0^\infty F_t(n)\, dn$$

or

$$\sigma_y^2 = <\overline{v'^2}>_t\, t^2,$$

where the symbol $< >_t$ indicates that the single-point Lagrangian velocity, v', is to be subjected to a moving average over the time t prior to computation of the variance.

Statistical methods for characterising diffusion using σ's are quite versatile but can be expected to lose accuracy with increasing time or distance of travel. They work best for horizontal diffusion because turbulence is usually far more homogeneous in the horizontal. For vertical diffusion, statistical measures give good results for short range ($x < 1$ km) but degrade rapidly as vertical turbulence varies greatly with height. Also as the plume grows, it moves into regions of substantially different turbulence statistics from those associated with the ground (Briggs, 1988).

3.3 Gaussian plume model

The normal or Gaussian distribution function provides a fundamental solution to the Fickian diffusion equation, and has been assumed as a continuous source diffusion model by many workers (see Hanna, 1982; Anon., 1989). Combination of the Gaussian assumption with an expression for the mean square particle diffusion, such as

$$\sigma_y^2 = \overline{y^2} = 2Kt$$

$$\sigma_y^2 = \sigma_v^2 t^2$$

or

$$\sigma_y = \sigma_v t f_1(t/T_L),$$

where f_1 is a universal diffusion function (Gifford, 1987), forms the basis of most of the practical plume diffusion formulae.

Strictly speaking, the Gaussian diffusion model applies only in the limit of large diffusion time and for homogeneous, stationary conditions, for which the diffusion problem may be stated in the form of the simple Fickian differential equation. Nevertheless, as Batchelor conjectured, the Gaussian function may provide a general description of average plume diffusion because of the essential random nature of this phenomenon, by analogy with the central limit theorem of statistics (Slade, 1968; Pasquill, 1974).

Assuming an instantaneous point source of material diffusing in three dimensions, with source strength Q in, say, grams, the concentration χ (or, as it is sometimes written, q) in grams per cubic metre is a function of x, y, z, the usual Cartesian coordinates, and time (see Figure 3.2). It is normally assumed that the source is at a fixed origin and t is the time of travel of the cloud from its instantaneous release. If σ^2 is the variance of the distribution of the dispersing cloud and it is assumed that the diffusion is isotropic, the Gaussian formula for an instantaneous point source is:

$$\chi(x,y,z,t) = \frac{Q}{(2\pi\ \sigma^2)^{3/2}} \exp\left(\frac{-r^2}{2\sigma^2}\right) \tag{3.16}$$

where

$$r^2 = (x - \bar{u}\ t)^2 + y^2 + z^2.$$

In the atmosphere, of course, diffusion is non-isotropic and under these conditions the Gaussian solution becomes

$$\chi(x,y,z,t) = \frac{Q}{(2\pi)^{3/2}\sigma_x\sigma_y\sigma_z} \exp\left(-\left(\frac{(x-\bar{u}\ t)^2}{2\sigma_x^2} + \frac{y^2}{2\sigma_y^2} + \frac{z^2}{2\sigma_z^2}\right)\right). \tag{3.17}$$

The case of a stack emitting pollutant may be approached by considering the continuous plume as a superposition of a large number of puffs. That is, the plume is regarded as resulting from the addition of an infinite number of overlapping average puffs, carried along the x axis by the mean wind, $\bar{u}$. Each puff is in reality composed of the average over an ensemble of puffs which have diffused for a time t and consequently have reached the position $(x, 0, 0)$. Mathematically this corresponds to integration of equation 3.17 with respect to t from 0 to ∞. This integration is not convenient, however, as the values of σ depend on t and hence x because

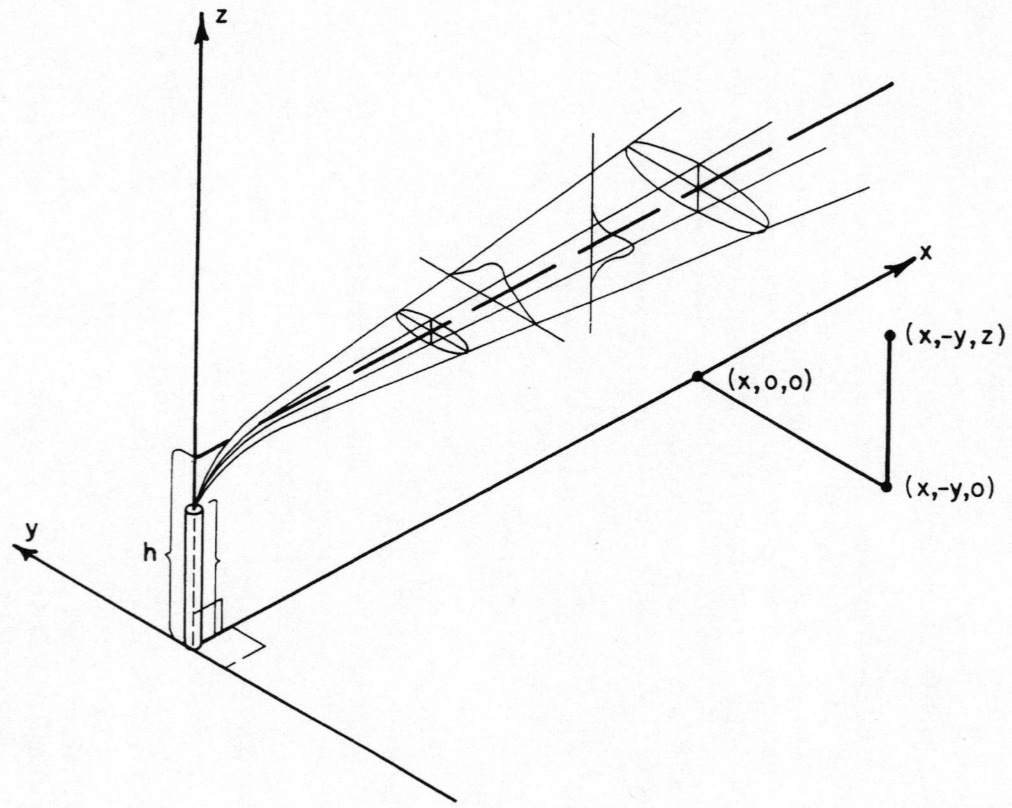

Figure 3.2 Coordinate system showing Gaussian distributions in the horizontal and vertical (after Turner, 1967)

$x = \bar{u}\, t$. As a practical matter, diffusion along the x axis is neglected by comparison with the gross transport along the x axis by the mean wind (Slade, 1968). With this simplification integration of equation 3.17 yields,

$$\bar{\chi}(x,y,z) = \frac{Q}{2\pi\, \sigma_y\, \sigma_z\, \bar{u}} \exp\left(-\left(\frac{y^2}{2\sigma_y^{\,2}} + \frac{z^2}{2\sigma_z^{\,2}}\right)\right), \tag{3.18}$$

where Q is now the continuous source strength, ordinarily in grams per second, and σ_y, σ_z can be regarded as functions of x.

Since most isolated continuous sources are located at or near the earth's surface, it is necessary to account for the presence of the physical barrier at the ground. This is usually done by assuming an imaginary image source located below the ground and symmetrical, with respect to the ground plane, to the actual source. This second 'image' source effectively produces a 'reflection' of material about the ground plane (see Figure 3.3). Hence the approach is only correct in the case of passive materials not absorbed or otherwise removed at the earth's surface. The resulting expression with reflection is:

$$\bar{\chi} = \frac{Q}{(2\pi\sigma_y\, \sigma_z\, \bar{u})} \exp\left(-\frac{y^2}{2\sigma_y^{\,2}}\right)\left(\exp\left(-\frac{(z-h)^2}{2\sigma_z^{\,2}}\right) + \exp\left(-\frac{(z+h)^2}{2\sigma_z^{\,2}}\right)\right), \tag{3.19}$$

where h is the elevation of the source above the ground plane. The elevation of the source is generally taken as the height at which the plume centreline becomes essentially level; that is the sum of the physical stack height and the added plume rise.

For the important case of ground-level concentrations this equation reduces to

$$\bar{\chi} = \frac{Q}{\pi\, \sigma_y\, \sigma_z\, \bar{u}} \exp\left(-\left(\frac{y^2}{2\sigma_y^{\,2}} + \frac{h^2}{2\sigma_z^{\,2}}\right)\right).$$

As this Gaussian model described by equations 3.17 to 3.19 forms the basis of many dispersion studies, it is important to note the implicit assumptions which are:

(i) Continuous emission from the source or emission times equal to or greater than travel times to the downwind position of interest, so that diffusion in the direction of transport can be neglected. That is, a plume-diffusion formula assumes that the release and sampling times are long compared with travel time from source to receptor. If the

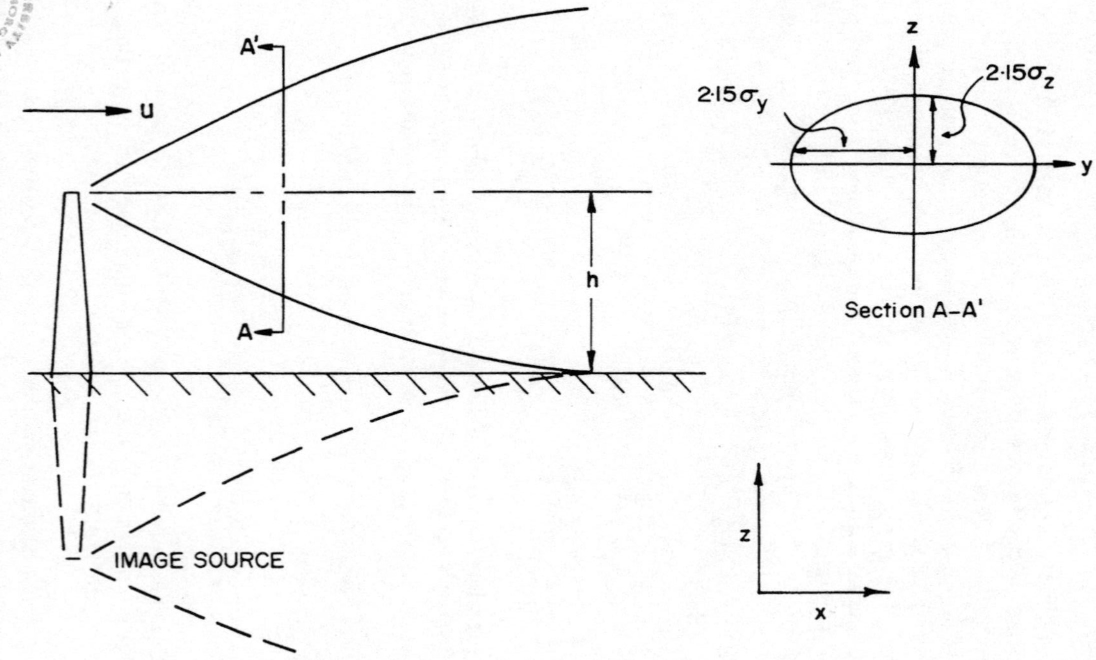

Figure 3.3 Diagram of a typical plume illustrating important concepts in the Gaussian plume formula (after Hanna *et al.*, 1982)

release or sampling time is short compared to the travel time then we are considering an instantaneous puff (equation 3.17) and cannot neglect diffusion in the direction of travel. This represents the essential difference between plume diffusion and puff diffusion.

(ii) The material diffused is a stable gas or aerosol (less than 20 μ diameter), which remains suspended in the air over long periods of time.

(iii) The equation of continuity holds. That is, none of the material emitted is removed from the plume and there is complete reflection at the ground.

(iv) Except where specifically mentioned, the plume constituents are assumed to have a Gaussian distribution in both the cross-wind and vertical directions.

(v) The Gaussian approach assumes steady-state conditions during the time interval for which the model is used, usually one hour. Of course during rapidly changing meteorological conditions, such as the passage of a front or the arrival of a sea-breeze, this assumption does not hold.

(vi) A constant wind speed, $\bar{u}$, is assumed. However, we have seen that wind speed increases with height near the surface. Hence for a moderate to strong vertical wind shear this assumption may introduce a considerable error. Furthermore, when the wind speed is variable, so no mean wind direction can be specified, and when the wind speed approaches zero, so the denominator in equation 3.18 approaches zero, the model cannot be applied.

(vii) Wind direction is assumed constant with height. However, the wind direction does change with height (the Ekman spiral), especially under stable conditions. Such a change has two important effects: an increase in the rate of spread and a deviation of the ground-level trajectory from a straight line. Secondly, the surface-wind direction in the xy plane is also assumed constant. This is a reasonable assumption for a uniform mesoscale area under steady conditions. However, hills and valleys have a profound influence on the surface-wind direction and tend to channel the flow.

(viii) The wind-shear effect on horizontal diffusion is not considered. This is a good approximation over short distances, but it becomes significant at distances greater than 10 km (Zib, 1977).

(ix) The dispersion parameters σ_y and σ_z are assumed to be independent of z and functions of x (and hence $\bar{u}$) alone. However, the eddy diffusivity increases with height near the surface. Ragland and Dennis (1975) note that when $\bar{u}$, σ_y or σ_z are considered independent of height, boundary-layer flow in the first several hundred metres may not be simulated. In addition σ_y and σ_z are functions of surface roughness (Golder, 1972), so that for varying terrain (for example, when the plume crosses a lake shore) they are not constant (cf. Kamst and Lyons, 1982b).

(x) The averaging time of all quantities ($\bar{u}$, σ_y, σ_z, χ) is the same.

Some of the above deficiencies are overcome by the Gaussian 'puff' model in which the plume from a continuous source is modelled by a series of puffs. This allows the wind speed and wind direction to vary in a horizontal direction, using a grid network, although these quantities are still considered to be independent of height (see, for example, Start and Wendell, 1974).

The critical input parameters to the Gaussian model are the effective stack height, h, and the variances σ_y and σ_z. These variances may be obtained by either theoretical or empirical approaches.

The American Meteorological Society Workshop on Stability Classification Schemes and Sigma Curves (Hanna *et al.*, 1977) noted that the following quantities are required to characterise σ_y and σ_z in the boundary layer:

(a) roughness length, z_0, and friction velocity, u_*, as measures of the mechanical turbulence;
(b) mixing depth, z_i, and Monin–Obukhov length, L, or the heat flux, H, as measures of the convective turbulence during the daytime; and
(c) wind speed, u, and the standard deviation of the wind direction fluctuations, σ_θ. The wind vector is needed to specify the transport wind, and σ_θ is required to estimate σ_y, especially in stable conditions.

It is strongly recommended that turbulence measurements be used to measure these quantities, but still many studies estimate diffusion based on the simpler Pasquill stability classes because they are easier to use and tend to give satisfactory results. More importantly, the instruments, time and expertise required to gain higher quality turbulence observations cannot always be justified. Nevertheless, it must always be borne in mind that the sigma curves based on the Pasquill classes are experimental observations and only strictly applicable to the area of their derivation. In other words, they are not appropriate in instances such as complex terrain and for distances greater than 10 kilometres, where either direct turbulence measurements or theoretical extrapolations are necessary.

The scheme, proposed by Pasquill (1961) involves subjective estimates of the structure of turbulence in the atmospheric boundary layer. He obtained I, the vertical spreading, and N, the lateral spreading, from routine meteorological observations, consisting of wind speed, insolation and cloudiness. He presented this information about I and N as functions of six atmospheric stability classes, designated A to F, where A corresponds to extremely unstable, F to extremely stable, while D corresponds to neutral conditions (see Table 3.1). Gifford (1961) modified the scheme somewhat

Table 3.1 Meteorological conditions defining Pasquill turbulence types

Surface wind-speed ms^{-1}	Daytime insolation			Nightime conditions	
	Strong	Moderate	Slight	Thin overcast or ≥ 4/8 low cloud	Cloudiness ≤ 3/8
< 2	*A*	*A–B*	*B*	–	–
2–3	*A–B*	*B*	*C*	*E*	*F*
3–4	*B*	*B–C*	*C*	*D*	*E*
4–6	*C*	*C–D*	*D*	*D*	*D*
> 6	*C*	*D*	*D*	*D*	*D*

Notes: *A*, extremely unstable; *B*, moderately unstable; *C*, slightly unstable; *D*, neutral (applicable to heavy overcast day or night); *E*, slightly stable; *F*, moderately stable; (for A–B take average of values for *A* and *B*, etc.)

Source: after Pasquill, 1961

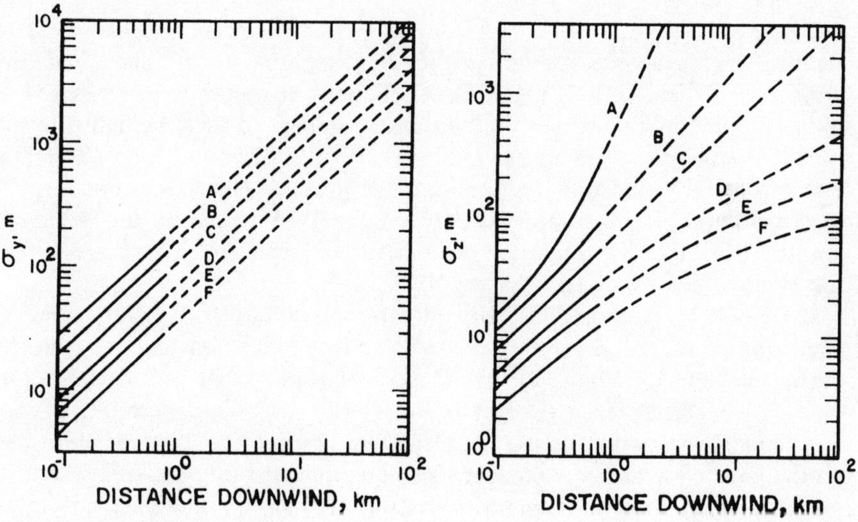

Figure 3.4 Pasquill–Gifford curves

by converting I and N into families of curves of $\sigma_y(x)$ and $\sigma_z(x)$ respectively. These are called the Pasquill–Gifford (PG) curves. They are based on a sampling time of three minutes and a roughness length of 3 cm for ground level releases with wind speeds > 2 ms^{-1} (see Figure 3.4).

Although these classes and the corresponding curves are widely used, they are most effective for vertical diffusion from surface and near-surface sources and reasonably effective for lateral diffusion over short averaging times. This diffusion is caused by small turbulent eddies which are not influenced by terrain slopes or surface inhomogeneities as much as the larger eddies (Briggs, 1988).

Singer and Smith (1966) put forward another scheme, the so-called Brookhaven National Laboratory system (BNL), which is based on data taken over a period of fifteen years and refers to plumes released at 108 m. The sampling times for the dispersion parameters were of the order of an hour. The BNL 'stable' curve falls distinctly below the PG '*F*' curve, despite the fact that it applies to a much rougher surface ($z_0 \approx 1$ m). It is most likely that this is because of the rapid decrease of turbulence with height (Hanna *et al.*, 1978); in addition, equivalence of the stability classifications between the two schemes (PG and BNL) is in some doubt (Pasquill, 1976a).

Carpenter *et al.* (1971) summarised twenty years' dispersion data representing helicopter sampling of SO_2 emitted from stacks of the Tennessee Valley Authority (TVA). The lapse-rate values were measured at plume height, in contrast to the two other studies, so that the data showed no super-adiabatic temperature gradients, though during daytime there is a high probability that these existed near the ground. Hence it is not surprising to see significant differences between the TVA results on the one hand and the PG and BNL results on the other hand. The TVA data have sampling times of 2–5 minutes (Weber, 1976).

Briggs (1973) proposed a series of interpolation formulae for σ_y and σ_z curves based on the above three schemes. Most weight was given to the PG curves at short ranges with the BNL curves given most weight at intermediate distances. At larger distances the TVA data were used. They are intended primarily for use in calculating ground-level concentrations, in particular the maximum values of these quantities for plumes from elevated stack sources, so that they reflect diffusion data for a higher source at greater downwind distances. These curves are summarised in Table 3.2.

Weil and Brower (1984) redefined the A–D dispersion curves in terms of u/w_* to yield

$$\sigma_y = [0.08^2 + (0.56w_*/u)^2]^{0.5}x,$$

$$\sigma_z = [0.06^2 + (0.56w_*/u)^2]^{0.5}x.$$

Table 3.2 Formulas recommended by Briggs (1973) for $\sigma_y(x)$, $\sigma_z(x)$ where $10^2 < x < 10^4$ m

Pasquill	σ_y(m)	σ_z(m)
	Open-country conditions	
A	$0.22\,x\,(1 + 0.0001\,x)^{-\frac{1}{2}}$	$0.20\,x$
B	$0.16\,x\,(1 + 0.0001\,x)^{-\frac{1}{2}}$	$0.12\,x$
C	$0.11\,x\,(1 + 0.0001\,x)^{-\frac{1}{2}}$	$0.08\,x\,(1 + 0.0002\,x)^{-\frac{1}{2}}$
D	$0.08\,x\,(1 + 0.0001\,x)^{-\frac{1}{2}}$	$0.06\,x\,(1 + 0.0015\,x)^{-\frac{1}{2}}$
E	$0.06\,x\,(1 + 0.0001\,x)^{-\frac{1}{2}}$	$0.03\,x\,(1 + 0.0003\,x)^{-1}$
F	$0.04\,x\,(1 + 0.0001\,x)^{-\frac{1}{2}}$	$0.016x\,(1 + 0.0003\,x)^{-1}$
	Urban conditions	
A–B	$0.32\,x\,(1 + 0.0004\,\mathrm{x})^{-\frac{1}{2}}$	$0.24\,x\,(+ 0.001\,x)^{\frac{1}{2}}$
C	$0.22\,x\,(1 + 0.0004\,\mathrm{x})^{-\frac{1}{2}}$	$0.20\,x$
D	$0.16\,x\,(1 + 0.0004\,\mathrm{x})^{-\frac{1}{2}}$	$0.14\,x\,(1 + 0.0003\,x)^{-\frac{1}{2}}$
E–F	$0.11\,x\,(1 + 0.0004\,\mathrm{x})^{-\frac{1}{2}}$	$0.08\,x\,(1 + 0.0015\,x)^{-\frac{1}{2}}$

Several studies have been carried out over various terrains to obtain σ_y and σ_z; comparisons have been made with the above results. For instance, Miller (1978) compared data obtained over rough terrain (z_0 about 100 cm) with both the PG and Briggs interpolation curves. He found that σ_y and σ_z for elevated releases over rough terrain in general were larger than predicted by either the PG or Briggs curves and that surface roughness has a greater effect on σ_z than σ_y.

Maas and Harrison (1977) found that σ_y at distances greater than 1 km from a natural seep, acting as a point source, were lower than the PG values for class *F*. They attribute this to the lower level of mechanically generated turbulence over water and the consequent production of only small-scale eddies.

Davison *et al.* (1977) found that in the Athabasca River Valley, Canada, values for σ_y and σ_z were consistently higher than the PG curves. The σ_z values did not increase with distance as quickly as predicted, perhaps due to elevated inversions. Correspondingly the centre path concentrations were found to be low. The increased mixing was thought to be due to the rougher surface. They showed that the σ_y and σ_z discrepancies attributable to turbulent mixing are approximately equal, which gives support to the idea of enhanced three-dimensional mixing. They further suggest that the assignment of a single stability class for a boundary layer, a layer with several different stabilities in the vertical is not appropriate.

Golder (1972) using micro-meteorological data, calculated *L* (the Monin–Obukhov length) values and Pasquill stability classes and obtained curves

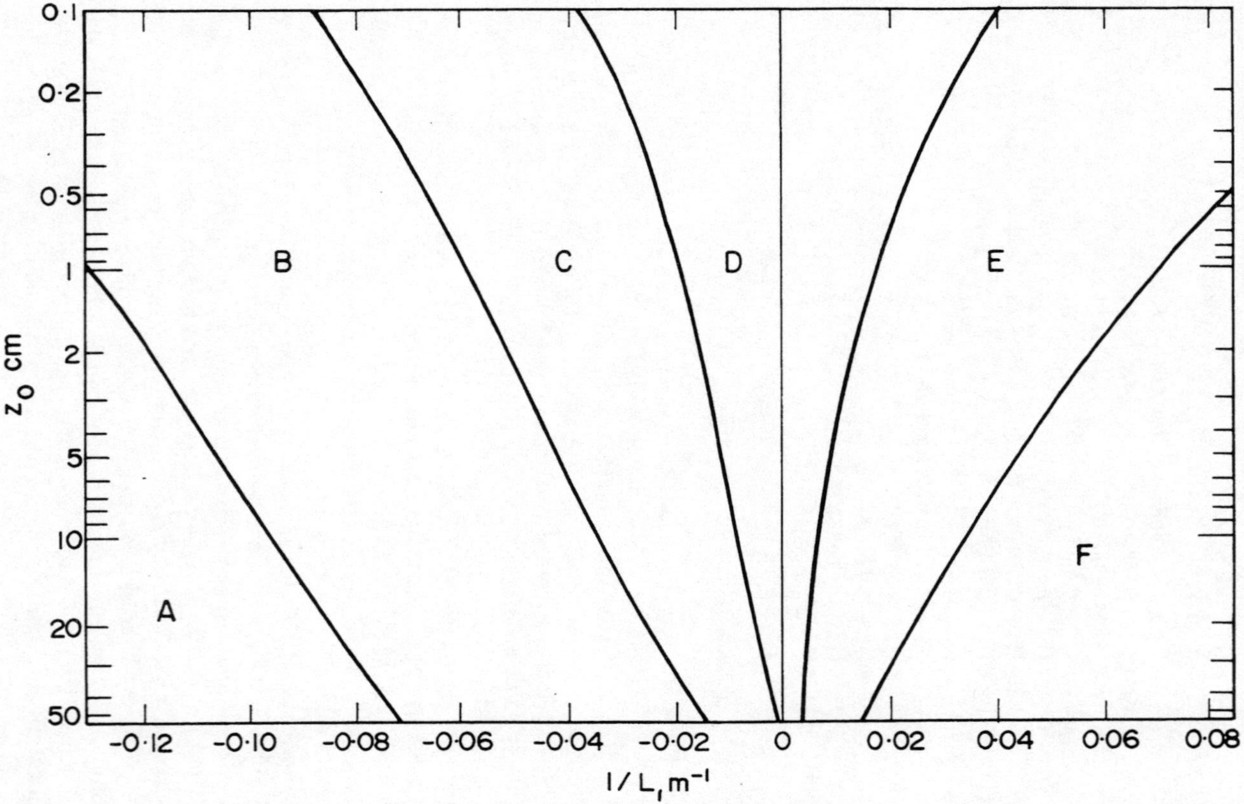

Figure 3.5 Curves showing Pasquill's stability types as a function of the Monin–Obukhov length scale and the aerodynamic roughness (after Golder, 1972). Reprinted by permission of Kluwer Academic Publishers.

showing Pasquill's turbulence types as functions of L and roughness length (see Figure 3.5). Although the qualitative stability categories correspond generally to direct measurements of L and the intensity of boundary-layer turbulence, there is considerable scatter.

Luna and Church (1971) obtained Richardson numbers for levels at 3 m and 40 m and established the relevant Pasquill turbulence classes. By comparing the two sets of results, they found that the categories are established in the proper sequence, but that the neutral class D does not seem to be an indicator of zero lapse rate of potential temperature (neutral condition).

Atwater and Londergan (1985) noted that the stability class estimate inferred from the observed dispersion often differed by two or more classes from the estimate based on meteorology. They also noted that the observed horizontal and vertical dispersion classes often differed from one another by two or more classes. On the other hand, Draxler (1987) observed that the Pasquill–Gifford dispersion curves with the stability category determined from the wind direction fluctuation performed well over Washington, DC, whereas the Briggs dispersion curves resulted in considerable undercalculation especially at night.

Although such experiments highlight the difficulty of specifying the turbulent mixing within the atmosphere, the analysis of dispersion experiments is also dependent on the assumptions of conservation of tracer mass (that is no loss of the chemical tracer through deposition, chemical reaction, etc.) and an assumed vertical profile. Gryning *et al.* (1983) reanalysed the Praire Grass experiment and showed that deposition and vegetative uptake of SO_2 lead to a 20–25 per cent reduction in the ground-level concentration at 200 m from the source. Hence failure to account for this pollutant loss would imply larger values of σ_z than are occurring.

In keeping with the recommendations of Hanna *et al.* (1977), Pasquill (1974, 1976b) and Draxler (1976) have shown that irrespective of surface roughness and stability

$$\sigma_y = \sigma_\theta \, x \, \mathrm{f}(x),$$

where σ_θ is the measured standard deviation of the horizontal wind direction and $\mathrm{f}(x)$ is a 'universal' correction factor, values of which are shown in Table 3.3. At distances greater than 20 kilometres, Pasquill (1976b) recommends incorporating another term to include the total change in mean wind direction over the vertical depth of the plume, $\Delta\theta$. That is,

$$\sigma_y = (\sigma_\theta^2 \, x^2 \, f^2(x) + 0.3 \, \Delta\theta^2 \, x^2)^{1/2}.$$

In choosing the appropriate σ_z curve, Hanna *et al.* (1977) recommended that the elevation of the source and roughness of the site be considered in

Table 3.3 Correction factor for σ_y curves

x (km)	0.0	0.1	0.2	0.4	1	2	4	10	> 10
$f(x)$	1.0	0.8	0.7	0.65	0.6	0.5	0.4	0.33	$0.33\,(10/x)^{1/2}$

Source: after Pasquill, 1976b.

relation to the data base in deriving the original curves. To this end, Gifford (1976) has summarised guides available for converting between the various schemes. Nevertheless, it should not be assumed that the plume thickness exceeds the depth of the mixed layer and hence, if the Gaussian model is used, σ_z should be held constant beyond $\sigma_z = 0.8\ z_i$ (Hanna *et al.*, 1977). Current sigma schemes have been extensively reviewed by Irwin (1983), who notes that the applicability of a particular scheme is dependent on source height, surrounding roughness and stability.

Briggs (1985) showed that the Praire Grass observations for the convective boundary layer could be represented as

$$\sigma_y/z_i = 0.6X$$

for $X < 1$ and

$$\sigma_y/z_i = 0.6X/(1 + 2X)^{0.5}$$

for $X > 1$, where

$$X = (w_* x)/(z_i u)$$

is a non-dimensional distance that can also be thought of as a non-dimensional time as it is defined as the ratio of the travel time x/u to the convective time scale z_i/w_*. At large distances (or times) the equation reduces to $\sigma_y/z_i = 0.4X^{0.5}$ in accord with equation 3.14.

A convenient function that satisfies both the short and long time limits of statistical theory is

$$f_1 = \frac{1}{(1 + 0.5t/T_L)^{0.5}}$$

and Weil (1988b) used this in defining σ_y as

$$\sigma_y = \sigma_v t\ f_1.$$

Over larger diffusion times and distances out to 1,800 km, Carras and Williams (1988) noted that the assumption of a Gaussian distribution is inconsistent with their observations. In particular, they suggest an accelerated diffusion regime where σ_y^2 is proportional to t^3 for $t > 3$ hours (Gifford, 1982, 1984) and the domination of quasi two-dimensional turbulence for $t > 15$ hours. This non-Gaussian diffusion is also evident in the analysis of the long-range dispersion from the Chernobyl accident (ApSimon *et al.*, 1989).

Despite the simple formulation of the Gaussian models, they have been used widely and often with good results. Although assuming K_z constant with height, they do not assume it constant with time or travel distance (see for discussion, Venkatram and Wyngaard (1988); Rayner (1987)). That is, the Gaussian formulation is able to correctly specify an increase in diffusivity as the plume grows, because of the larger range of eddies acting to disperse the plume.

The Gaussian models usually give reasonable estimates of pollutant concentration levels, but their use is limited to other than light wind or variable conditions. The dispersion parameters σ_y and σ_z show considerable scatter in the studies described above. This is attributable to the different surface roughnesses. The Gaussian models should work reasonably well for areas with constant surface roughness by making σ_y and σ_z dependent on them. However, for areas with varying surface roughness, when the Gaussian model cannot cope, alternative methods must be found.

3.4 Plume rise

Before discussing the application of the Gaussian model, the other critical input parameter to consider is the effective stack height. This is the height at which the plume becomes essentially level and it will rarely correspond to the physical height of the stack. In fact, plume rise can increase the effective stack height by a factor twice to ten times the physical stack height and is thus an important factor in determining ground level concentrations.

As with the sigma curves, the literature abounds with empirical formulations for estimating plume rise as well as more recent work based on the fundamental laws of fluid mechanics; that is, the conservation of mass, potential density and momentum (Weil, 1988a). Many early studies concentrated on fitting regression equations and hence have produced highly empirical formulations that are of limited applicability. Some of the more common ones are summarised in Table 3.4. Since the plume rise from a stack generally occurs over some distance downwind these equations should not be applied within the first few hundred metres of the stack. Of course, from our discussion of the Lagrangian velocity correlation coefficient, it is clear that the Gaussian model is not applicable close to the source where you have short diffusion times.

The fundamental break from these empirical formulations was due to the

Table 3.4 Early empirical plume-rise formulations

Author	Expression	Remarks
Holland	$\Delta h = \frac{V_s d}{u} (1.5 + 2.68 \times 10^{-3} p (\frac{T_s - T_a}{T_s}) d)$	Developed from large sources (stack diameter 1.7–4 m, stack temperature 82–204°C)
Concawe	$\Delta h = \frac{5.53 Q_h^{1/2}}{u^{3/4}}$	Regression equation – large buoyant plumes
Lucas–Moore–Spurr	$\Delta h = \frac{135 Q_h^{1/2}}{u}$	Regression equation based on work of Priestly
Rauch	$\Delta h = \frac{47.2 Q_h^{1/2}}{u}$	Same as Lucas *et al.*, but different data base
Moses and Carson	$\Delta h = \frac{A}{u} (-0.029 V_s d + 5.53 Q_h^{1/2})$	Regression equation many different data sources

Notes:
Δh : plume rise (m)
V_s : stack exit velocity (ms^{-1})
d : inside stack diameter (m)
u : wind speed (ms^{-1})
p : atmospheric pressure (hPa)
T_s : stack gas temperature (K)
T_a : air temperature (K)
Q_h : heat emission rate (k cal s^{-1})
A : coefficient dependent on stability

Stability	A
Unstable	2.65
Neutral	1.08
Stable	0.68

Source: after Carson and Moses, 1969.

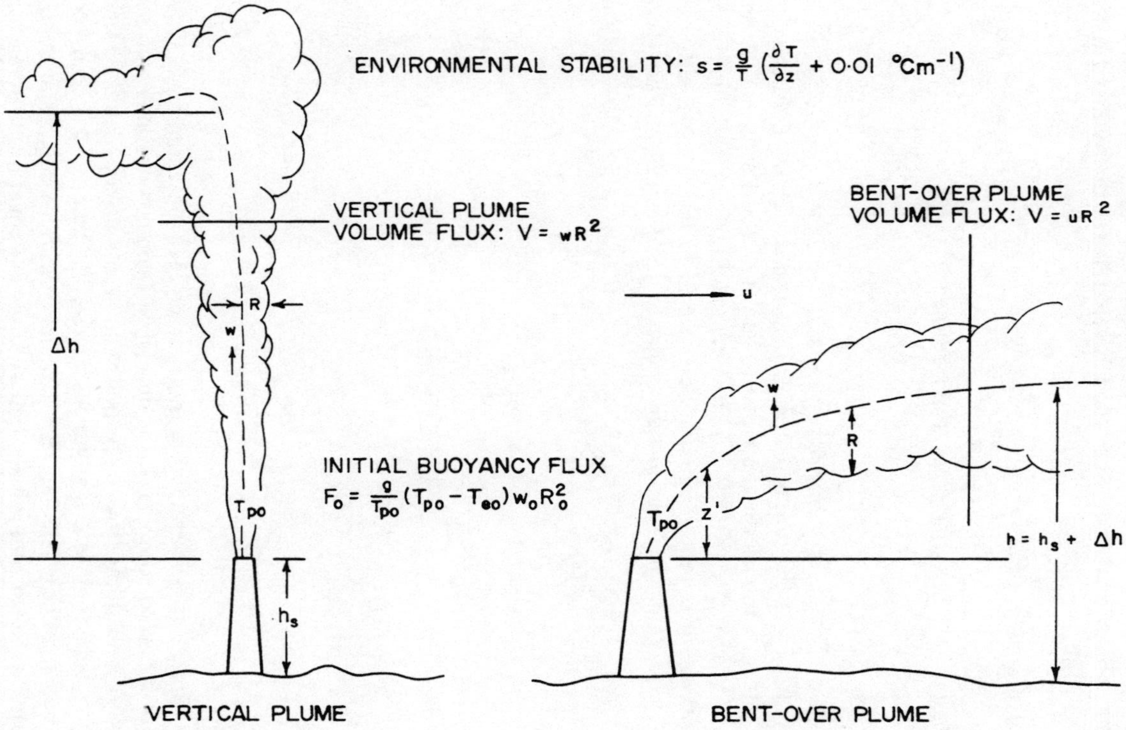

Figure 3.6 Schematic diagram of vertical and bent-over plumes which illustrate some of the parameters and variables used in plume-rise calculations (after Hanna, 1982; Hanna *et al.*, 1982)

work of Briggs (1975) and his plume-rise equations are now used in most studies. Figure 3.6 represents a schematic drawing of a vertical and a bent-over plume, which illustrates the major variables and parameters used in estimating plume rise.

From a consideration of the fluid dynamics, Briggs (1975) was able to show that for a bent-over buoyancy-driven plume, the plume trajectory takes the form

$$z = 1.6\, F_0^{1/3}\, u^{-1}\, x^{2/3} \tag{3.20}$$

where F_0 is the initial buoyancy flux, defined as

$$F_0 = w_0\, R_0^2\, \frac{g}{T_{p0}}\, (T_{p0} - T_{e0}),$$

where w_0 (ms^{-1}) is the initial plume vertical speed, R_0 (m) the inside stack radius, g (ms^{-2}) the acceleration due to gravity, T_{p0} (K) the initial plume temperature and T_{e0} (K) the ambient temperature at stack height. The coefficient 1.6 in equation 3.20 can be expected to be accurate within ± 40 per cent with variations due to downwash or local terrain effects (Briggs, 1984).

Final plume rise for a bent-over plume in stable conditions is given by

$$\Delta h = 2.6 \left(\frac{F_0}{uS} \right)^{1/3}, \tag{3.21}$$

where S is the ambient stability parameter defined as

$$S = \left(\frac{g}{T} \right) \left(\frac{\partial T}{\partial z} + \Gamma_d \right)$$

$$= \left(\frac{g}{T} \right) \frac{\partial \theta}{\partial z},$$

which is similar to the stability parameter defined by equation 2.37. The coefficient 2.6 in equation 3.21 was developed from comparisons with many observations, including the TVA power-plant plume-rise data. For these data, the coefficient is slightly conservative in that it will lead to an underestimate of the plume rise (Hanna *et al.*, 1982). The ambient temperature gradient and wind speed required by equation 3.21 represents the average values over the expected height range of the plume. In the case of wind speed this can be estimated from a measured wind speed at 10 m (the standard height for anemometers) by

Table 3.5 Parameter p for estimating the wind speed as a function of stability and surface type

Stability class	Urban	Rural
A	0.15	0.07
B	0.15	0.07
C	0.20	0.10
D	0.25	0.15
E	0.40	0.35
F	0.60	0.55

Source: after Irwin, 1979

$$u(z) = u_{10} \left(\frac{z}{10} \right)^p ,$$

where the parameter p is a function of stability class as shown in Table 3.5. At heights above 200 m the value of wind speed calculated at 200 m should be used.

In the case of calm winds, a buoyant plume will rise vertically until (Briggs, 1981)

$$\Delta h = 5.3 \; F_0^{1/2} \; s^{-3/8} - 6 \; R_0. \tag{3.22}$$

The correction term '6 R_0' says that a virtual source exists six stack radii below the actual stack height.

Under neutral conditions the stability parameter will approach zero and Briggs recommends the following formula to estimate plume rise

$$\Delta h = 1.54 \left(\frac{F_0}{u \; u_*^2} \right)^{2/3} h_s^{1/3}, \tag{3.23}$$

where u_* is the friction velocity and h_s is the physical stack height.

All of the above equations for estimating plume rise assume that the plume is not influenced by a nearby building or other obstacle. Many industrial stacks are built within a factory complex and the plume will be influenced by air flow around other buildings in the complex or the stack itself (Hosker, 1984). The influence of mechanical turbulence around a building or stack can significantly alter the effective stack height. This is especially true with high winds, when the beneficial effect of high stack-gas velocity is at a minimum and the plume is emitted nearly horizontally.

Low pressure in the wake of a stack may cause the plume to be drawn downward behind the stack to form downwash. Downwash can be prevented by maintaining the efflux velocity (w_0) greater than the cross-wind velocity (u). In fact, it is generally recognised that downwash will not

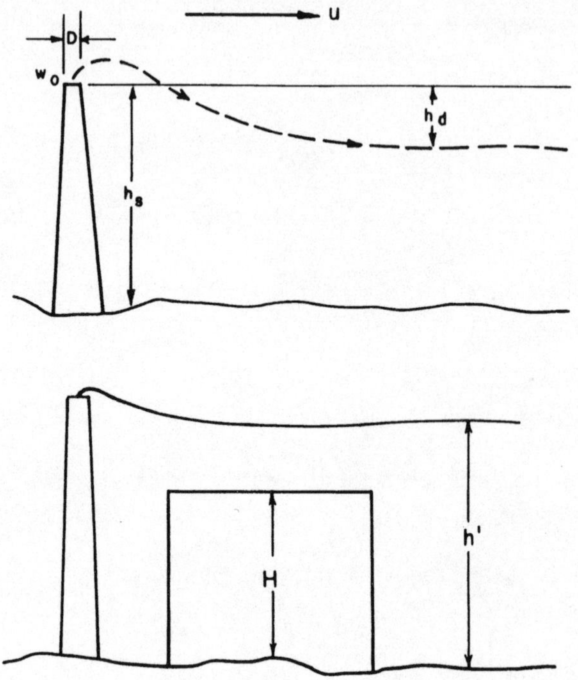

Figure 3.7 Stack downwash and downwash due to a nearby building (after Briggs, 1973)

occur for $(w_0/u) > 1.5$ (Hanna *et al.*, 1982). For values of this ratio less than 1.5, Briggs (1973) suggests that the distance (h_d) the plume downwashes below the top of the stack is given by

$$h_d = 2\left(\frac{w_o}{u} - 1.5\right) d, \tag{3.24}$$

where d is the internal stack diameter.

Downwash due to the presence of nearby buildings depends to a large extent on the individual site and is best evaluated through wind-tunnel studies. A region of disturbed flow surrounds an isolated building, generally to at least twice its height and extends downwind five to ten times its height. Obviously it is important to keep the plume away from this wake cavity, where it would be brought to the ground and recirculated in a region with low ventilation. Building the stack 2.5 times the height of the highest building adjacent to the stack will usually overcome these effects.

For plumes emitted near the building wake cavity, Briggs (1973) recommends that stack downwash is calculated using equation 3.24. The effective stack height after downwash is then (see Figure 3.7)

$$h' = h_s - h_d.$$

Next, ξ is defined as the smaller of W, the cross-wind width of the building, and H, the height of the building, and if h' is greater than H + 1.5 ξ, then the plume is out of the wake and the effective plume height equals h'. Otherwise the plume is affected by the building. Thus Briggs (1973) recommends the effective plume height (h_e) be calculated as follows:

for $h' > H + 1.5\ \xi$,

$$h_e = h'; \tag{3.25}$$

for $H + 1.5\ \xi > h' > H$,

$$h_e = 2\ h' - (H + 1.5\ \xi); \tag{3.26}$$

for $h' < H$,

$$h_e = h' - 1.5\ \xi; \tag{3.27}$$

The plume is assumed to be trapped in the cavity if the calculated value of h_e is less than 0.5 ξ. In this case the plume can be modelled as a ground level source of initial area ξ^2.

3.5 Applications of the Gaussian model

Despite its limitations, the Gaussian model has enjoyed a wide degree of popularity mainly due to its simplicity and the fact that it gives reasonable results. It forms the basis of regulatory air-quality modelling in Australia (Sawford and Ross, 1985) and elsewhere. Various authors have modified the standard equation (3.18, 3.19) to extend its range of application (see Appendix for listing of standard air-quality models), even to incorporating enhanced dispersion in building wakes (for example, Huber, 1988).

For example, a stable layer aloft above an unstable layer will restrict the vertical diffusion of pollutant. Obviously, once the plume reaches the stable layer we can no longer assume a Gaussian distribution in the vertical as the stable layer will prevent further diffusion aloft and in fact will tend to reflect the pollutant. Such an effect can be incorporated in equation 3.18 if we assume the height of the base of the stable layer is L. At the height of 2.15 σ_z, above the plume centre-line, the concentration in the plume is one-tenth the plume centre-line concentration at the same distance. This is normally taken as the physical extent of the plume and hence when one-tenth of the plume centre-line concentration extends to the stable layer, it is reasonable to assume that the plume is starting to be affected by the stable layer. Until this time we can assume that the plume has a Gaussian distribution in the vertical. Assuming that this distance is x_L downwind from the stack, beyond this distance the plume will be reflected from the stable layer, leading to increased concentrations between the stable layer and the ground. At a distance of 2 x_L downwind the plume is assumed to be completely mixed between the stable layer and the ground; that is, there is no variation in concentration with height, although we still assume a Gaussian distribution in the horizontal. Equation 3.18 reduces to

$$\chi = \frac{Q}{\sqrt{2\pi}\ \sigma_y\ L\ u} \exp\left\{-\ ½ \left(\frac{y}{\sigma_y}\right)^2\right\}. \tag{3.28}$$

Concentrations between x_L and $2x_L$ can be estimated by interpolation between the standard Gaussian estimate at x_L and the modified Gaussian estimate (equation 3.28) at $2x_L$.

A similar modification can be made to equation 3.18 to account for the limitation of horizontal diffusion imposed by a valley wall. Generally this will occur when the valley width approximates to 2 σ_y. In this case one would assume uniform mixing in the horizontal at twice this distance downwind and maintain the Gaussian distribution in the vertical. Under these conditions equation 3.19 for a ground level concentration would be reduced to

$$\chi = \frac{2Q}{\sqrt{2\pi}\ \sigma_z\ Y\ u} \exp\left\{-\ ½ \left(\frac{h}{\sigma_z}\right)^2\right\}, \tag{3.29}$$

where Y is the width of the valley.

To investigate the pollution concentrations due to a highway, it is necessary to assume that the highway consists of an infinite line of point sources. In other words, if we assume that the wind is blowing along the x axis at right angles to the highway which then lies along the y axis, we can approximate the effect of the highway by integrating the Gaussian equation for a point source with respect to y. Ground-level concentrations downwind of a continuously emitting infinite line source, when the wind direction is normal to the line, can then be expressed as

$$\chi = \left(\frac{2}{\pi}\right)^{½} \frac{q}{\sigma_z\ u} \exp\left\{-\ ½ \left(\frac{h}{\sigma_z}\right)^2\right\}, \tag{3.30}$$

where q is the source strength per unit distance, such as g s^{-1} m^{-1}. Since the line source is infinite in the y direction, the horizontal dispersion parameter σ_y does not appear in the equation.

If the wind is not perpendicular to the line source but rather blowing at an angle ϕ to it, then equation 3.30 becomes

$$\chi = \left(\frac{2}{\pi}\right)^{1/2} \frac{q}{\sigma_z\, u \sin\phi} \exp\left\{-\tfrac{1}{2}\left(\frac{h}{\sigma_z}\right)^2\right\}. \tag{3.31}$$

It is apparent that if the wind is blowing parallel to the line source this equation will predict zero concentration on either side of the line source. Such a condition is unrealistic as obviously there will be some horizontal diffusion of pollutant from the line that was ignored in deriving equation 3.30 because the line source is oriented along the y axis and not the x axis. Hence equation 3.31 should not be used for $\phi < 45°$, as diffusion in the cross-wind direction cannot be neglected in this instance.

A final important modification to the Gaussian model is for the case of fumigation. This occurs when a surface-based inversion is eliminated by the upward transfer of sensible heat from the ground surface, and can occur when either the ground is heated by solar radiation or when air flows from over a cold surface to a relatively warm surface. Such a situation occurs locally with the sea breeze wherein the air passes from over the cool ocean surface to over the warm land surface (see Figure 3.8). In this case pollutants emitted into the stable layer will be mixed vertically downward when they are caught by the thermal eddies from the heated land surface below. Unlike transient fumigation events associated with the erosion of nocturnal inversions, shore-line fumigation events may persist for several hours (Kerman *et al.*, 1982). As various meteorological and source parameters vary this can lead to peak concentrations spread over a large area and consequently has generated considerable interest (Lyons, 1975).

To estimate ground-level concentrations under fumigation one assumes that the plume was initially emitted into stable air and the appropriate values of σ_y and σ_z for stable air are used. The ground-level concentration is then given by

$$\chi_F = \frac{Q\left(\int_{-\infty}^{p} \frac{1}{\sqrt{2\pi}} \exp(-0.5\,p^2)\,dp\right)}{\sqrt{2\pi}\,\sigma_{yF}\,u\,h_i} \exp\left\{-\tfrac{1}{2}\left(\frac{y}{\sigma_{yF}}\right)^2\right\}, \tag{3.32}$$

where the integral in brackets accounts for the portion of the plume that is mixed downward by the thermal eddies (Turner, 1967). This equation assumes that the thermal mixing extends to h_i and that $p = (h_i - h)/\sigma_z$. Obviously if the thermal mixing extends up to the effective stack height, half of the plume will be mixed downward, the other half remaining in the stable air aloft.

When the fumigation concentration is near its maximum, equation 3.32 can be approximated by

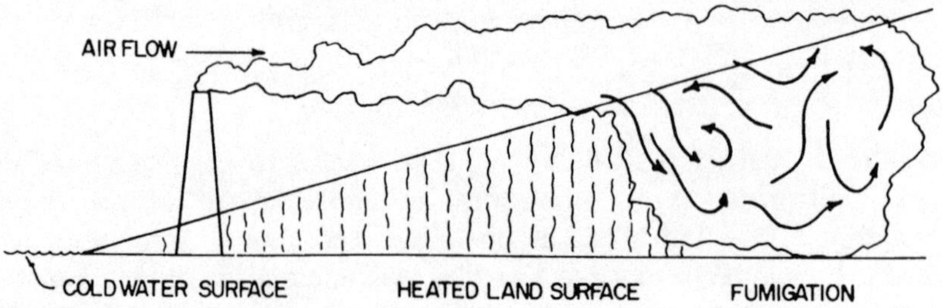

Figure 3.8 Fumigation during a sea breeze

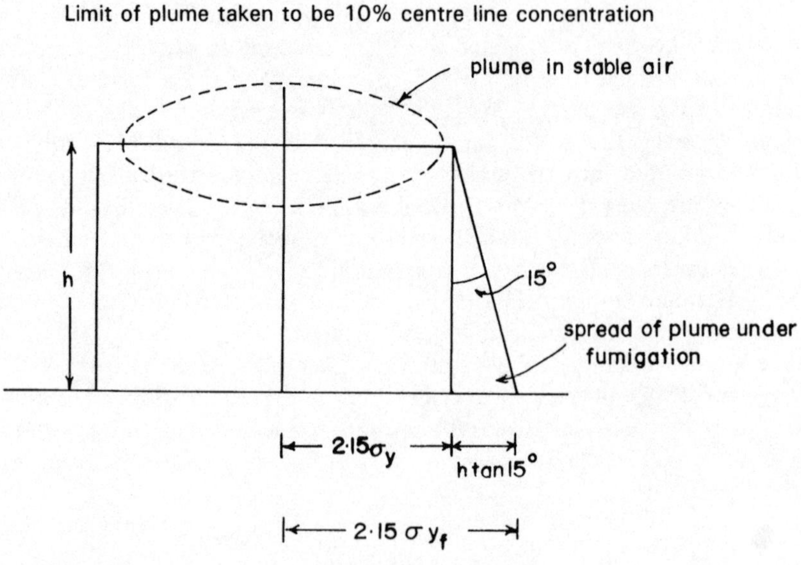

Figure 3.9 Horizontal spread of plume under fumigation

$$\chi = \frac{Q}{\sqrt{2\pi}\ u\ \sigma_{yF}\ h_i} \exp\left\{-\ ½ \left(\frac{y}{\sigma_{yF}}\right)^2\right\}, \tag{3.33}$$

$$h_i = h + 2\ \sigma_z.$$

The main difficulty with the use of equations 3.32 and 3.33 lies in estimating a realistic value for the horizontal diffusion since when the plume is mixed downward by the thermal eddies, some additional horizontal spreading will occur. Bierly and Hewson (1962) overcame this by noting that the edge of the plume spreads outward at an angle of approximately 15° and suggested that the σ_{yF} for fumigation equals the σ_y for stable conditions plus one-eighth the effective height of emission as shown in Figure 3.9. That is

$$\sigma_{yF} = \frac{2.15\ \sigma_y\ (\text{stable}) + h \tan 15^\circ}{2.15}$$

$$= \sigma_y\ (\text{stable}) + \frac{h}{8}.$$

More recent models of shoreline fumigation follow the same general principles but have alternate schemes for handling the amount of plume trapped in the thermal internal boundary layer (Venkatram, 1988b).

3.6 Models based on K-theory

The K-theory models are based on equation 3.5

$$\frac{dq}{dt} = \frac{\partial q}{\partial t} + u\frac{\partial q}{\partial x} + v\frac{\partial q}{\partial y} + w\frac{\partial q}{\partial z}$$

$$= \frac{\partial}{\partial x}\left(K_x\frac{\partial q}{\partial x}\right) + \frac{\partial}{\partial y}\left(K_y\frac{\partial q}{\partial y}\right) + \frac{\partial}{\partial z}\left(K_z\frac{\partial q}{\partial z}\right), \quad (3.5)$$

which assumes that the gradient-transfer hypothesis is valid. There are some theoretical objections to the use of this equation. Corrsin (1974) states that, in a boundary layer simulated in the laboratory, the length- and time-scales of the transporting action should be sufficiently uniform and small compared with the length- and time-scales of variation in the mean field gradients of the property undergoing transport. He further notes that from the basic similarities of neutrally stratified boundary layers in both laboratory and atmosphere it would be surprising if the above conditions were not violated in the atmosphere as well. Despite this objection the method is known to give useful results for momentum transfer in the atmospheric boundary layer (Pasquill, 1975).

Another theoretical objection is that the plume cross-section affected by eddies larger than itself cannot be described by the gradient-transfer relation. For a stable atmosphere, however, the eddies are small, especially when decoupling of the surface layer occurs, so that the gradient-transfer relation can justifiably be used (Kamst and Lyons, 1982b).

The flexibility of the *K*-model may be appreciated by realising that K_x,

K_y and K_z can be specified as a function not only of stability, but also of surface roughness, space and time. In this case solutions cannot be found analytically, so numerical methods must be used.

Shir and Shieh (1974) solved equation 3.5 letting $K_x = K_y = 500\ m^2\ s^{-1}$ be independent of space and time. They note that the horizontal diffusivity has no significant effect on the results and its behaviour is not well understood, especially under stable conditions. For K_z they used Shir's (1973) turbulence transport model. Near the surface u_* and K_z were calculated from the Monin–Obukhov theory using an appropriate value for z_0. They found that the model was especially satisfactory for both light and strong wind conditions and an improvement over the Gaussian model.

MacCracken *et al.* (1978) developed an air-quality model on the basis of equation 3.5 for the San Francisco area, a topographically complex terrain. Since poor air quality usually occurs during inversion conditions and little is known concerning the vertical structure of the atmosphere below the inversion, they chose to simplify the situation by treating the atmosphere below the inversion base as a single layer when calculating the horizontal transport. They obtained realistic K_z profiles under these conditions. In addition they used variable K_x and K_y values. It was found that the results of this model, LIRAQ 1, indicated generally good agreement in magnitude, temporal phasing and spatial patterns where allowance was made for grid size, spatial domain and unrepresentativeness of suburban observation stations in primarily rural areas (Duewer *et al.*, 1978).

The two models mentioned above require considerable input data and computer time. Most other models introduce simplifications by neglecting one or more terms in equation 3.5. It is usually assumed that

$$\frac{\partial}{\partial z}\left(K_z \frac{\partial q}{\partial z}\right) >> w \frac{\partial q}{\partial z}$$ vertical diffusion is much greater than vertical advection

$$u \frac{\partial q}{\partial x} >> \frac{\partial}{\partial x}\left(K_x \frac{\partial q}{\partial x}\right),$$ horizontal advection is much greater than horizontal diffusion, and

K_y is independent of position.

This yields a simplified version of equation 3.5:

$$\frac{\partial q}{\partial t} + u \frac{\partial q}{\partial x} + v \frac{\partial q}{\partial y} = \frac{\partial}{\partial z}\left(K_z \frac{\partial q}{\partial z}\right) + K_y \frac{\partial^2 q}{\partial y^2}. \qquad (3.34)$$

Tangerman (1978) used equation 3.34 to obtain concentration values. She obtained vertical profiles of u, v and K_z as functions of surface roughness, geostrophic wind, stability and the Coriolis parameter using a numerical boundary-layer model by Fiedler (1972). It was assumed that

$K_y = 2\,K_z^{\max}$, where $K_z^{\max}$ is the maximum value for K_z in the boundary layer. She found that the pollutant concentration distribution is strongly determined by the intensities of the eddy diffusivities and the roughness of the terrain, indicating the general importance of correct specification of these parameters.

Other models neglect changes in direction, so that the pollutant is advected only along the x axis, that is,

$$\frac{\partial q}{\partial t} + u\,\frac{\partial q}{\partial x} = \frac{\partial}{\partial y}\left(K_y\,\frac{\partial q}{\partial y}\right) + \frac{\partial q}{\partial z}\left(K_z\,\frac{\partial q}{\partial z}\right). \tag{3.35}$$

Ragland and Dennis (1975) used this equation and compared results with those predicted by the Gaussian model. They used the Monin–Obukhov theory in the surface layer to obtain u, u_* and K_M (assuming a given z_0) and extended u and K_M into the boundary layer using u_*. Their results indicate that realistic wind and diffusivity profiles yield a greater percentage of pollutant mass closer to the ground than predicted by the Gaussian model, primarily because of lower wind speeds and diffusivities near the ground. The maximum ground-level concentration for a given set of meteorological conditions was not found to differ much from the Gaussian result in either magnitude or position. They also note the importance of correctly specifying z_0.

Maul (1978) reduced equation 3.5 further by assuming steady-state conditions

$$\left(\frac{\partial q}{\partial t} = 0\right)$$

and ignoring cross-wind variations

$$\left(\frac{\partial}{\partial y}\left(K_y\,\frac{\partial q}{\partial z}\right) = 0\right):$$

thus obtaining,

$$u\,\frac{\partial q}{\partial x} = \frac{\partial}{\partial z}\left(K_z\,\frac{\partial q}{\partial z}\right). \tag{3.36}$$

The advantage of this model is that analytical solutions can be obtained, however, it is restricted in its applications. Maul (1978) found that the exact form of K_z and u profiles is not important, provided the overall features are reproduced. This model is somewhat more flexible than the Gaussian model in that it will accept vertical profiles of K_z and u.

The advantages of the K-model are that realistic conditions (three-dimensional variations in the wind field and diffusivity field) can be simulated. Appropriate simplifications may be introduced by neglecting one or more terms. A disadvantage of this model is that it does not allow for an increase in K_z with plume travel time, by virtue of the increase in size of the plume. Thus, near the source dispersion is overestimated to some extent, although under stable conditions this is minimised, because of the small range of eddies.

Hanna and Paine (1987) and Gryning *et al.* (1987) defined dispersion modelling in terms of the meteorological scaling parameters. In particular, Gryning *et al.* (1987) presented different models for the vertical dispersion under each scaling regime to account for the different scaling of the turbulence (see section 2.10). This allows for non-Gaussian dispersion in the vertical although the lateral dispersion is constrained to be Gaussian and the models are only applicable for horizontally homogeneous conditions within ten km of the source.

3.7 Other models

Details of the modelling process have been explored at length at conferences and workshops and in numerous reports of field studies. These consider the practical evaluation of the standard deviation with the statistical approach, Gaussian plume modelling and *K*-theory (for example, Venkatram and Wyngaard, 1988; KAMS (1982)). Extreme non-gradient effects simulated using the so-called higher order closure modelling of diffusion in the atmosphere have received much attention. In this approach, the gradient-transfer approximation is not made, but the higher moment equations are used. For example, in second order closure, the second moment equation expressing the total time derivative of the flux in terms of other second and third moments of the turbulence fluctuations is used. The task is to make its solution possible by appropriate 'modelling' of these other terms (Pasquill, 1975; Stull, 1988). Results have been obtained by Llewellen and Teske (1976a, 1976b), Donaldson (1973) and Yamada and Mellor (1975), amongst others, and the approximations required to close the system of equations reviewed by Moeng and Wyngaard (1986, 1989).

Numerical mesoscale models have been used to specify physically consistent flow fields (Pielke *et al.*, 1983) and pollutant dispersion incorporated within these through the use of random walk diffusion techniques (Reid, 1979; McNider, 1983; Thomson, 1984, 1985). Such models overcome many of the limitations of Gaussian theory but require extensive computational facilities and have not yet been accepted for regulatory applications.

An alternative approach to solving atmospheric diffusion problems is provided by the pseudospectral and orthogonal collocation methods (Wengle *et al.*, 1978; Fleischer and Worley, 1978). They essentially use the

K-theory and solve equation 3.5. Although good agreement with other models is found, there appears to be a substantial number of difficulties to be overcome, such as numerical difficulties and the use of various aperiodic boundary conditions. Such models do not appear to have as much potential as the higher order closure models.

Apart from theoretical approaches based on the conservation of mass equation, several empirical models exist, which are site specific. Smith and Jeffrey (1973), for example, proposed an empirical equation to estimate daily averages of sulphur dioxide. The review by Smith (1984) presents the current inadequacy of air-quality models. It is saddening but none appear up-to-date and there is little basis for choosing a model due to uncertainties in the diffusion process and inadequate input data.

3.8 Removal mechanisms

The standard Gaussian plume description of atmospheric dispersion does not account for the removal, transformation and resuspension of pollutant. In fact we have assumed a stable, passive, conservative pollutant in our discussions so far. Nevertheless, various complicated processes act to transform atmospheric pollutants, removing them from the diffusing cloud, deposit them on the earth's surface, and also pick them up again once deposited – that is, to resuspend them. Hence removal, transformation and resuspension processes must be incorporated in our description of pollutant dispersal.

Radioactive decay of a species may be accounted for according to the exponential decay scheme

$$\overline{\chi}_R = \overline{\chi} \exp(-\lambda_R t), \tag{3.37}$$

where $\overline{\chi}$ is the quantity of the radioactive species present in the absence of decay, calculated from the standard Gaussian equation, and λ_R is the fraction of the species that decays per unit time. One half of the nuclei of the species will decay in a time

$$T_R = \frac{0.693}{\lambda_R},$$

where T_R is the half life of the species.

As a practical matter equation 3.37 may also be used in approximating some simple non-radioactive chemical transformations (see Chapter 4, residence time).

Air pollutants often consist of particles with a large range of sizes and shapes, ranging from aerosols to large particles several hundred micrometres across. Particles or droplets at the larger end of this range fall fairly quickly from the plume and Van der Hoven (1968) has suggested that for particle radii greater than 100 μm the particles are falling so fast that diffusion is no longer important. In this case, a simple ballistics approach based on the mean wind speed and the particle size will enable calculation of the particle trajectory.

Particles or droplets under this size will settle following their gravitational settling speed adjusted for eddy diffusion. The diffusion of these particles can be accounted for by a simple modification of the plume formula, the so-called tilted plume model, which consists in replacing h (the effective height) in the Gaussian equation by the quantity $(h - xv_g/\bar{u})$. The resulting formula describes the behaviour of large particles from a stack plume reasonably well, and has been used in studies of the behaviour of 'drift' droplets, the large droplets that are formed in cooling towers and carried out in their plumes (Gifford, 1975). Under this model the plume centre-line will strike the ground at $x = h\bar{u}/v_g$ and roughly half of the material will be deposited by this distance. This model conserves mass and according to Horst (1980) is probably a reasonable approximation for $(v_g x/uh) < 1$.

A more general treatment of plume deposition is afforded by a modification to the plume equation, which assumes that the rate at which material is deposited on the surface, ω (g s^{-1} m^{-2}), is proportional to the airborne concentration close to the ground. That is,

$$\omega = v_d \, \bar{\chi} \, (x, y, 0), \tag{3.38}$$

where the proportionality parameter, v_d, has the dimensions of velocity, ms^{-1}, and is called the deposition velocity. In general the deposition velocity is not equal to the particle's gravitational settling speed and must be determined empirically. Provided the concentration profile remains constant with distance downwind from the source, the deposition velocity will not be a function of distance and equation 3.38 can be used to predict the dry deposition of gases and small particles.

The magnitude of the deposition velocity can also be defined as

$$|v_d| = (r_a + r_b + r_c)^{-1}$$

where r_a is the aerodynamic resistance (common to all gases) between a specified height and the surface, r_b is the quasi-laminar sublayer resistance and r_c is the bulk surface resistance (Wesely, 1989; Baldocchi *et al.*, 1987). Typical deposition velocities for gases are given by Dasch (1986, 1987) and Shanley (1989), whereas Nicholson (1988b) reviewed measurements for particles. Arritt *et al.* (1988) have also suggested that variations in both land use and terrain can induce regions of enhanced deposition implying changes in the deposition velocity as a function of surface characteristics.

The major problem with this concept lies in accounting for the effect of the deposited materials on the plume concentration $\chi(x, y, z)$. This problem has been attacked by two distinct procedures, the source-depletion and the surface-depletion models. Source-depletion consists in calculating the total deposition at each downwind distance and subtracting this from the source. That is, the apparent strength of the source is allowed to vary with downwind distance to account for the diminishing amount of material remaining aloft. Surface-depletion on the other hand represents the deposition flux as a material sink, that is, a negative source for diffusion downwind of the point of deposition. This is the more realistic procedure as it permits the form of the cloud concentration function χ to be modified by deposition but it tends to be computationally complex. If the source is depleted by deposition

$$Q = Q(x)$$

$$\text{and } \frac{\partial Q}{\partial x} = -\int_{-\infty}^{\infty} \omega(x, y)\, dy$$

$$= -\left(\frac{2}{\pi}\right)^{1/2} \frac{v_d Q}{u\,\sigma_z} \exp\left(\frac{-h^2}{2\sigma_z^2}\right)$$

which leads to

$$\frac{Q(x)}{Q(0)} = \{\exp \int_0^x \frac{dz}{\sigma_z \exp\,(h^2/2\sigma_z^2)}\} \tag{3.39}$$

where $\alpha = -(2/\pi)^{1/2}\,(v_d/u)$.

Thus the flux to the surface can be determined by integration, given a suitable expression for σ_z. In common with all source depletion models, this artificially assumes that deposition is a loss at the source rather than the surface, even though it correctly evaluates the deposition in terms of the airborne concentration near the surface (Horst, 1980). Thus depletion occurs over the whole depth of the plume rather than only at the surface.

An alternative source depletion model was developed by Overcamp (1976) in which the deposition is accounted for by reducing the strength of the conventional image source. That is

$$\chi = \frac{Q}{2\pi\sigma_y\sigma_z u} \exp\left(\frac{-y^2}{2\sigma_y^2}\right)\left(\exp\left\{\frac{-(z - h + (v_g x/u))^2}{2\sigma_z^2}\right\}\right.$$

$$\left. + \alpha(x_G) \exp\left\{\frac{-(z + h - (v_g x/u))^2}{2\sigma_z^2}\right\}\right), \tag{3.40}$$

where the reflection coefficient $\alpha(x_G)$ can be computed by solving an implicit relation for x_G

$$\{h - \frac{v_g x_G}{u}\} \frac{\sigma_z(x)}{\sigma_z(x_G)} = z + h - \frac{v_g x}{u}$$

and

$$\alpha(x) = 1 - \frac{2v_d}{v_g + v_d + v_t},$$

where v_t is a turbulent diffusion velocity defined by Csanady (1955) as the rate at which the Gaussian plume expands in the vertical. That is,

$$v_t = \frac{(u\,h - v_g\,x)}{\sigma_z} \frac{d\,\sigma_z}{dx}.$$

Overcamp's (1976) model only reduces the image source rather than both real and image sources and thus allows deposition to remove material from the lower portions of the plume; this begins to show a non-Gaussian vertical concentration distribution. As the image source accounts for complete reflection, any reduction in the concentration contributed by it is effectively deposited to the surface.

In reviewing diffusion–deposition models, Horst (1980) noted that Overcamp's model is the easiest to use and gives realistic results at short downwind distances. At large downwind distances, however, equation 3.39 is more accurate. Despite this, it must be realised that adequate data are not available to validate these models and their accuracy was only tested against a mathematically correct modification of the Gaussian plume model which in itself is empirical through its reliance on the values of σ_y and σ_z (Horst, 1980).

As well as dry deposition through gravitational settling, particles in the atmosphere can be removed through wet deposition, that is by the action of a falling droplet of water. Wet deposition is often divided into rainout (within cloud scavenging) and washout (below cloud scavenging) but the two processes can be modelled similarly (Hosker, 1980).

A falling droplet of water sweeps out an air volume and its path will intersect that of some of the pollutant particles in this volume. This fraction of particles in the droplet's path is known as the target efficiency, E and the fraction of these contacting particles that actually stick is the retention efficiency, R. Hence the actual collection efficiency, C, is

$$C = ER.$$

Very little is known of R and it is usually taken as 1 so that the washout coefficient is given by

$$\Lambda_w = \int F\ E\ A\ \mathrm{d}(D), \tag{3.41}$$

where D is the diameter of the scavenging droplets, A is the cross-sectional area of drops of diameter D, and F is the flux density of drops, that is, the drops per unit area, time and interval of diameter, $\mathrm{d}(D)$.

If the washout process occurs uniformly over the plume volume, then the plume concentration after washout is

$$\bar{\chi}_w = \bar{\chi} \exp\,(-\ \Lambda_w t).$$

This method is, strictly speaking, only applicable to particles of a single size and to highly reactive gases, which are irreversibly captured by the precipitation (Hanna *et al.*, 1982). The method can only be applied to aerosols with a range of diameters if an empirical value of Λ_w is available for particles of that type and size.

Other methods of modelling wet removal use the washout ratio. If C_0 and χ_0 are the concentration of effluent in the precipitation (e.g. raindrops) and in the air respectively, at some reference height, then

$$W_r = \frac{C_0}{\chi_0}, \tag{3.42}$$

where W_r is the washout ratio (Misra *et al.*, 1985; Chan and Chung, 1986; Savoie *et al.*, 1987; Venkatram *et al.*, 1988). The flux of effluent to the surface as a result of the precipitation is

$$F_w = C_0\ J_0, \tag{3.43}$$

where J_0 is the equivalent rainfall rate in, for example, millimetres per hour. If W_r is known and χ_0 has been measured or estimated from a plume model, then

$$F_w = \chi_0\ W_r\ J_0. \tag{3.44}$$

At short distances from the stack, plume washout may nearly double local acid deposition (ten Brink *et al.*, 1988).

The washout ratio can also be used to define a wet deposition velocity by analogy to the dry deposition velocity. That is,

$$v_w = \frac{F_w}{\chi_0} = W_r\ J_0, \tag{3.45}$$

and this can be used in a similar way to develop models of the wet deposition process (Hanna *et al.*, 1982). Of course, it must be remembered that washout is not an irreversible process as gases washed out of a tall stack plume can be desorbed into the air below the plume.

3.9 Box models

An urban area contains thousands of individual sources ranging from home incinerators to large industrial complexes. Consequently, the direct application of a Gaussian diffusion model to each individual source is impractical and various simplifying assumptions have to be made. In general, contributions from small sources are combined into an area source with assumed strength in terms of mass pollutant per unit time per unit area that is taken as being constant across the area (Lyons *et al.*, 1990). Easily identified larger sources can still be treated individually and the concentrations at a receptor due to each source summed to determine the overall concentration.

If we assume that emissions in an urban area are constant over a distance Δx and that the pollutant is uniformly mixed between the ground and the mixing depth z_i (see Figure 3.10) then the continuity equation for this volume is (Hanna *et al.*, 1982)

$$\Delta x \, z_i \frac{\partial \chi}{\partial t} = \Delta x Q_a + u \, z_i \, (\chi_b - \chi) + \Delta x \frac{\partial z_i}{\partial t} (\chi_a - \chi), \qquad (3.46)$$

where Q_a is the area source strength (mass per unit time per unit area), χ_b is the upwind background concentration, χ_a is the concentration above the mixing depth and the mixing depth z_i is allowed to vary with time. This equation considers that the total change in concentration results from the contribution from the source, horizontal advection, the growth of the mixed layer and vertical advection.

If steady-state conditions exist, that is

$$\frac{\partial \chi}{\partial t} = \frac{\partial z_i}{\partial t} = 0,$$

and the background concentration is zero, equation 3.46 reduces to

$$\chi = \frac{\Delta x Q_a}{z_i \, u}, \qquad (3.47)$$

which is the box model solution.

An alternative development of the box model, by Hanna (1971) and Gifford and Hanna (1973), assumes that the mixing depth of the pollutant

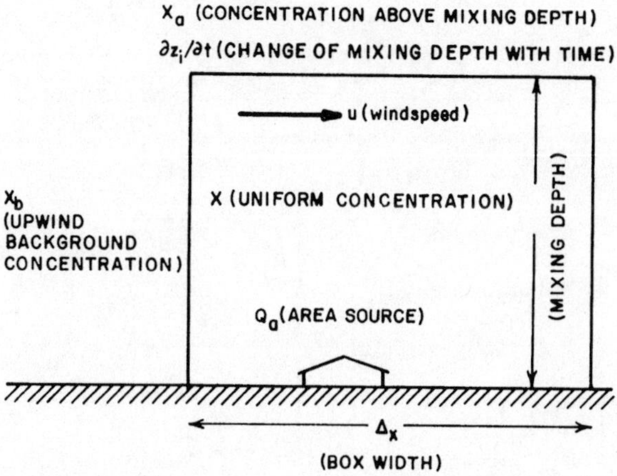

Figure 3.10 Parameters in box model (after Hanna *et al.*, 1982)

was proportional to the vertical dispersion parameter. They treat an area source as an infinite array of ground-level point sources of strength Q_a (x, y) and hence the concentration at the point (0, 0, 0) is

$$\chi = \int_0^\infty \int_{-\infty}^\infty \{\frac{Q_a}{\pi\ u\ \sigma_y\ \sigma_z} \exp\left(\frac{-y^2}{2\ \sigma_y^2}\right)\}\ dy\ dz.$$

Assuming that Q_a is a function of x alone this equation reduces to

$$\chi = \int_0^\infty \{\left(\frac{2}{\pi}\right)^{1/2} \frac{Q_a}{u\ \sigma_z}\}\ dx,$$

which can be written as

$$\chi = \left(\frac{2}{\pi}\right)^{1/2} \frac{\Delta x \bar{Q}}{u\ \bar{\sigma}_z}, \tag{3.48}$$

which is equivalent to equation 3.47.

The derivation of the box model and its application stems from the equation of continuity, which is a mass balance of pollutant in the box. A similar approach can be adopted in estimating indoor air pollution in a given volume (i.e. Wadden and Scheff, 1983; Ryan *et al.*, 1983, 1988).

For example, consider the system shown in Figure 3.11. In the overall air balance q is the volumetric flow rate for make-up air (q_o), recirculation (q_1), infiltration (q_2), exfiltration (q_3) and exhaust (q_4). Thus from a simple mass balance

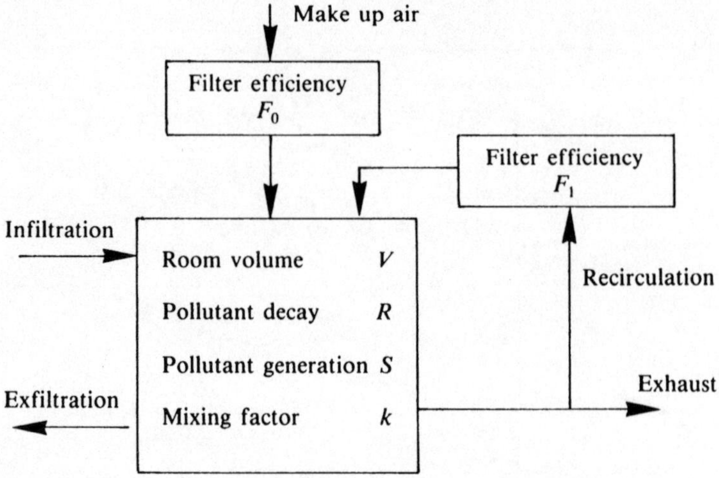

Figure 3.11 Indoor air quality model (after Shair and Heitner, 1974)

$$q_0 + q_2 = q_3 + q_4, \tag{3.49}$$

as the volume of air within the room remains constant. The pollutant mass balance leads to

$$V \frac{d\chi_i}{dt} = kq_0 \chi_0 (1 - F_0) + kq_1 \chi_i (1 - F_1)$$

$$+ kq_2 \chi_0 - k (q_0 + q_1 + q_2) X_i + S - R, \tag{3.50}$$

where χ_i is the indoor concentration, χ_0, the outdoor concentration; F is the filter efficiency for make-up (F_0) and recirculation air (F_1), V is the room volume, S is the indoor source emission rate, R is the indoor sink removal rate and k is a factor which accounts for inefficiency of mixing and is essentially the fraction of incoming air which completely mixes within the room volume.

The solution to equation 3.50 with boundary value $\chi_i = \chi_s$ at $t = 0$, is

$$\chi_i = \frac{k\{q_0(1 - F_0) + q_2\} \chi_0 + S - R}{km} \left(1 - \exp\left(- \frac{kmt}{V}\right)\right)$$

$$+ \chi_s \exp\left(- \frac{kmt}{V}\right), \tag{3.51}$$

where $m = q_0 + q_1 F_1 + q_2$; this equation has been used by a variety of authors in the analysis of indoor air pollution (i.e. Shair and Heitner, 1974; Selway *et al.*, 1980; Ishizu, 1980).

The major difficulty in using equation 3.51 is in estimating the infiltration q_2 and exfiltration q_3 which are functions of temperature and pressure differences between indoor and outdoor air (Wadden and Scheff, 1983). The static pressure, P_v, over a building surface can be described by

$$P_v = 0.6008\ v^2, \tag{3.52}$$

where v is the wind velocity (ms^{-1}) and P_v is in Pascals. The pressure difference due to a thermal gradient, ΔP_c, is given by

$$\Delta P_c = 0.0342\ Ph \left(\frac{1}{T_0} - \frac{1}{T_i} \right), \tag{3.53}$$

where P is atmospheric pressure (Pascals), T_0 is the outside temperature (K), T_i the inside temperature (K) and h is the distance from the neutral pressure level. If cracks and openings are uniformly distributed in the vertical direction, h will be half the building height. The total pressure drop across a wall on the windward side is $P_v + \Delta P_c$.

The flow resulting from this pressure difference is

$$q_2 \approx q_3 = K_F\ (\Delta P)^n, \tag{3.54}$$

where ΔP is the pressure between indoors and outdoors, K_F is a flow coefficient and n is an empirical exponent between 0.5 (turbulent flow) and 1.0 (laminar flow).

$$\Delta P = \frac{P_v + \Delta P_c}{1 + (A_w/A_L)^{1/n}} \tag{3.55}$$

where A_W, A_L are the leakage areas on the windward and leeward sides (Wadden and Scheff, 1983).

Equation 3.51 is similar in form to the solution of Lettau's (1970) simplification of equation 3.46 although obviously applied to a room rather than a city. Lettau defined the equilibrium box concentration given by equation 3.47 as χ^*. He also defined a scaling time $\Delta x/u$ as the flushing time or the time required for the air to pass completely over the urban area. Taking a non-dimensional time $t^* = tu/\Delta x$ and neglecting χ_b, χ_a Lettau (1970) reduced equation 3.46 to

$$\frac{\partial \chi}{\partial t^*} = \chi^* - \chi,$$

which has the solution

$$\chi = \chi^* + (\chi_s - \chi^*) \exp(-t^*) \tag{3.56}$$

where χ_s is the initial value of the concentration, that is the boundary value at $t = 0$.

As time increases the concentration will approach the equilibrium concentration given by equation 3.47 in the same manner as the concentration within a room will approach the steady state given by

$$\chi_i = \frac{k(q_0(1 - F_0) + q_2)\chi_0 + S - R}{km},$$

assuming that none of these parameters are functions of time.

Chapter 4

POLLUTANTS AND THEIR PROPERTIES

Our discussions so far have concentrated on the physical aspects of the transport, diffusion and removal of pollutants without considering the properties of an individual pollutant. For downwind distances of more than a few kilometres, one of the major uncertainties in dispersion calculations is the loss or transformation of individual pollutants through chemical reactions (Grefen and Löbel, 1987). To formulate chemically reactive dispersion models we must ideally account for all the reactions, reactants, products, reaction rates and meteorological conditions involved in a chemical system. Our current knowledge of atmospheric chemistry is insufficient to completely specify this entire system. Thus, the formulation of reactive models requires compromises, particularly in determining the important reactions to model and valid parameterizations of the reaction-rate phenomena.

4.1 Residence time and reaction rates

The residence time of a pollutant is the time required to empty a system or reservoir of that pollutant. That is,

$$\tau = \frac{M}{R}, \qquad (4.1)$$

where M is the mass of the pollutant in the system and R is the mass removed per unit time. Such a definition as equation 4.1 assumes there is no inflow of pollutant into the system. Obviously if there is an inflow, F, the system will not empty as quickly, or if F equals R we have a steady state in which there is neither depletion nor gain of material from/to the system.

If the reaction or removal rate is first order, that is,

$$R = k \cdot M$$

where k is the turnover constant, then from equation 4.1, τ is independent of M. In the case of no reverse reaction the turnover constant is more or less equivalent to the first order reaction rate constant for the chemical reaction

$$A \xrightarrow{k} B.$$

Then

$$\frac{dA}{dt} = -\,k\,A$$

and k is equal to the fraction loss per unit time. On integration

$$\int_{A_0}^{A} \frac{dA}{A} = -\int_0^t k\,dt$$

$$\frac{A}{A_0} = \exp\,(-\,k\,t)$$

$$= \exp\left(-\,\frac{t}{\tau}\right), \qquad (4.2)$$

where $\tau = 1/k$ is the time required for the original concentration A_0 to be depleted to 1/e, 0.378, of its original value.

The variation of constituents results from a large variety of effects including (i) source variation, (ii) wind variation, dispersion, turbulence, (iii) reactivity of the gas, and (iv) removal processes, all of which can be characterised by residence times.

For instance, SO_2 in the atmosphere is eventually converted to sulphate particles. In the absence of catalytic surfaces in the free atmosphere the depletion rate is less than 0.1 per cent hr^{-1}. This means that the turnover constant is 0.001 hr^{-1} and the residence time of SO_2 is roughly 1,000 hours or forty days. As can be seen, this residence time depends on the size of the reservoir and the flow rate from it. The reaction of SO_2 in the free atmosphere is quite slow. In water drops in clouds, fog or smog, the reaction is much faster and, in reasonable conditions, might consume 1 per cent of the SO_2 in an hour or have a residence time of four days.

The times need not be short. For instance, the atmosphere as we know it has evolved through the billions of years since the earth's formation. Initially, it probably was pure CO_2, but through photosynthesis the O_2 content has built up to its present level. It is estimated that the oxygen content is presently changing at less than 1 ppm in a hundred years. This corresponds to a residence time of 2×10^7 years! It is thus a permanent gas for all practical purposes. CO_2, on the other hand, due to the burning of fossil fuels, is currently increasing at a rate of more than 1 ppm per year. Since the CO_2 concentration is about 350 ppm, this corresponds to a turnover constant of 0.3 per cent yr^{-3} or a residence time of about 300 years (Ehhalt, 1987).

In the case of series reactions of the form:

$$A \xrightarrow{k_1} B \xrightarrow{k_2} C$$

the net effective value of k is given by

$$k = \frac{1}{\dfrac{1}{k_1} + \dfrac{1}{k_2}} \tag{4.3}$$

and, since rate constants generally cover about ten orders of magnitude, one expects that one of the two constants will be much larger than the other. Assuming

$$k_1 >> k_2,$$

then the first term in the denominator of equation 4.3 disappears and

$$k \approx k_2,$$

or the slowest reaction dominates. On the other hand, when parallel reactions occur:

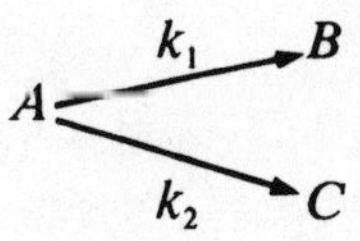

the overall kinetic constant is

$$k = k_1 + k_2$$

and the reaction with the largest k dominates. This, of course, is the

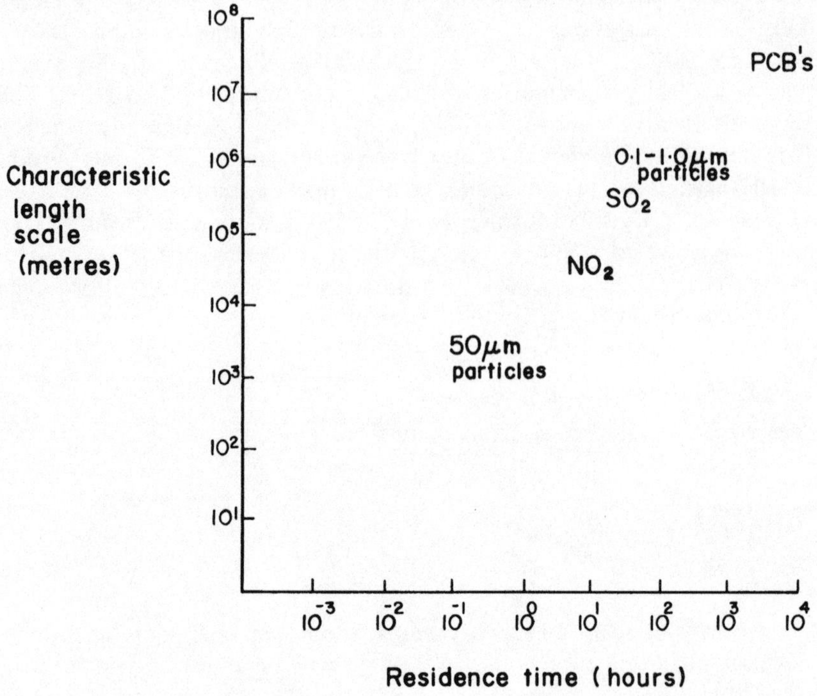

Figure 4.1 Estimated residence times for selected pollutant species and their associated horizontal transport scale. Estimates based on Junge (1972, 1974)

reaction that has the smallest τ. Most importantly we should recognise that τ depends on both the mass of material in the reservoir and the removal rate. That means that every reservoir, body or system has a characteristic residence time τ for each constituent (see Figure 4.1) though, of course, this τ may vary with concentration when the depletion is not a simple linear function of amount.

The residence time of atmospheric constituents can usually be associated with a length scale which is not so easily defined, but can be considered as a distance for depletion due to atmospheric diffusion, advection, or simple fallout. Figure 4.1 suggests a relationship between the two scales dependent on the size or chemical activity/inactivity.

When substances undergo chemical reactions, the reactions may be reversible. For example,

$$aA + bB \rightleftarrows cC.$$

The rate of forward reaction is proportional to the probability of finding a molecules of A and b molecules of B or

$$\text{forward rate } \alpha \underbrace{[A]\,[A]\ldots}_{a \text{ times}} \underbrace{[B]\,[B]\ldots}_{b \text{ times}}$$

where the bracket [ ] means 'concentration'

$$= k_f[A]^a\,[B]^b.$$

Likewise, for the reverse reaction:

$$\text{reverse rate } \alpha\ k_r[C]^c.$$

At equilibrium the forward rate is exactly equal to the reverse rate

$$k_f[A]^a\,[B]^b = k_r[C]^c.$$

This means that we can define an equilibrium constant

$$K = \frac{k_f}{k_r} = \frac{[C]^c}{[A]^a\,[B]^b}, \tag{4.4}$$

which forms what is called the Law of Mass Action. That is, if one increases the amount of A or B we expect an increase in C. Alternatively, if C is removed as it is produced, we expect more and more C to be formed. In this case, through the mass action of A and B, the concentration of C may be maintained. Hence to adequately model the transformation of a chemical species the concentrations of many species may need to be known even though it is likely that only a few reactions are important.

4.2 Sulphur compounds

Sulphur is an important pollutant species that occurs in the atmosphere primarily as SO_2, H_2S and particulate $SO_4^{=}$, though there are small amounts of mercaptans (Radical-SH) and methyl sulphide $(CH_3)_2S$. Natural sulphur comes from H_2S released by $SO_4^{=}$ reducing bacteria; anthropogenic sulphur is usually released as SO_2 from coal burning, petroleum combustion, and smelting processes. The H_2S is oxidized to SO_2 by an unknown process which has been generally ignored. Suffice it to say that though at release H_2S concentrations are about six times the SO_2 concentrations, the SO_2 concentrations in most regions of the atmosphere are comparable to the H_2S concentrations. The oxidation of H_2S is relatively fast, but not the SO_2 oxidation. In the best conditions, without catalytic influences, the SO_2 oxidation is only about 0.1 per cent per hour. This is thought to occur by one of the following mechanisms:

$$SO_2 + \begin{pmatrix} [O] \\ \text{atomic oxygen} \\ [OH] \\ \text{hydroxyl radical} \\ [HO_2] \\ \text{peroxy radical} \end{pmatrix} + \underset{\text{(any third body)}}{M} \rightarrow SO_3 + M^*.$$

The third body is necessary to absorb the excess energy and becomes activated during the reaction. The sulphur trioxide, SO_3, quickly takes on water becoming H_2SO_4 which in turn takes on water or a basic substance such as NH_3 to become sulphate ion, $SO_4^=$. The sulphate ion has a long lifetime in the atmosphere as particulate material of relatively small size ($<$ 0.1 μm). Even so, the amount of sulphur in sulphate, generally as ammonium sulphate, is normally less than the sulphur present as SO_2.

Sulphur dioxide has long been recognized as a major pollutant. It primarily comes from the burning of high sulphur fuels and the smelting of metal sulphides. All major air-pollution studies consider sulphur dioxide (see Freney and Nicolson, 1980; Bubenich, 1984; Galloway *et al.*, 1985; Graedel *et al.*, 1986; and Rodhe and Herrera, 1987). It is known that it can be conveyed thousands of kilometres (Nordo, 1976); the tall stacks of Sudbury in Canada and Mt Isa in Queensland, Australia, are but an indication of the major sources that allow conveyance of the gas over continental distances (Roberts and Williams, 1978). Still, there are regions of the earth relatively free of the effects (Murray, 1989). There is even a suggestion that in alkaline soils SO_2 emissions may increase crop yields but generally the effect is detrimental (Murray and Wilson, 1989). This gas also has a link with climatic change through the sulphate particles that are omnipresent in both the troposphere and the stratosphere. These are considered to affect a cooling of the troposphere (Wuebbles *et al.*, 1989).

The oxidation of SO_2 is by many different mechanisms including surfaces, gas-to-particle conversion and catalysts. The net effects have been simulated variously (Niewiadomski, 1989; Scott, 1984; and Scott and Hobbs, 1967). Here we consider only the simplest case of SO_2 reaction in water droplets. It should be remembered that SO_2 and NO_2 together are responsible for our global acid-rain problem. The reactions ultimately produce rains, fogs and clouds that have acidities greater than orange juice and sometimes approach a pH of 2.0 (Johnson and Gordon, 1987; Lefohn and Krupa, 1988). Sulphur dioxide oxidation is most rapid in the condensed phase, particularly in water drops or cloud droplets. The oxidation of the sulphite iron, $SO_3^=$, limits the conversion rate. That is, it is this reaction, of a series of reactions, that determines the net conversion rate; the reaction rate is not greatly affected by diffusional transport or other reactions. The SO_2 is transported close to the droplet by turbulent (large-scale) effects and brought to the droplet surface by molecular (small-scale) diffusion. Ammonia, NH_3, is similarly transported to the droplet. These gases dissolve in the droplet producing numerous ionic species, including the sulphite ion which is, in turn, oxidized to sulphate ion. The process is pH dependent as the amount of $SO_3^{=}$ becomes insignificant at pH values less than 5.

The details of the mechanism are as follows (Scott and Hobbs, 1967):

(i) The gases are in quasi-equilibrium with the liquid species

$$NH_4OH(s) \rightleftarrows NH_3(g) + H_2O(l) \qquad \frac{[NH_3][H_2O]}{[NH_4OH]} = \frac{P_{NH_3}}{[NH_4OH]} = K_{ha}$$

$$H_2SO_3(s) \rightleftarrows SO_2(g) + H_2O(l) \qquad \frac{[SO_2][H_2O]}{[H_2SO_3]} = \frac{P_{SO_2}}{[H_2SO_3]} = K_{hs}$$

$$H_2CO_3(s) \rightleftarrows CO_2(g) + H_2O(l) \qquad \frac{[CO_2][H_2O]}{[H_2CO_3]} = \frac{P_{CO_2}}{[H_2CO_3]} = K_{hc}$$

(ii) The liquid species are in quasi-equilibrium with the ionic species

$$NH_4OH \rightleftarrows NH_4^+ + OH^- \qquad \frac{[NH_4^+]\,[OH^-]}{[NH_4OH]} = K_a$$

$$H_2SO_3 \rightleftarrows H^+ + HSO_3^- \qquad \frac{[H^+]\,[HSO_3^-]}{[H_2SO_3]} = K_{1s}$$

$$HSO_3^- \rightleftarrows H^+ + SO_3^{=} \qquad \frac{[H^+]\,[SO_3^{=}]}{[HSO_3^-]} = K_{2s}$$

$$H_2CO_3 \rightleftarrows H^+ + HCO_3^- \qquad \frac{[H^+]\,[HCO_3^-]}{[H_2CO_3]} = K_{1c}$$

$$HCO_3^- \rightleftarrows H + CO_3^{=} \qquad \frac{[H^+]\,[CO_3^{=}]}{[HCO_3^-]} = K_{2c}$$

and

$$H_2O \rightleftarrows H^+ + OH^- \qquad \frac{[H^+]\,[OH^-]}{[H_2O]} \equiv [H^+]\,[OH] = K_w$$

where the square brackets should be interpreted to read 'the concentration of', using moles per litre or molar mixing ratios; $[H_2O]$ is taken by convention (standard state) to be unity as the molar mixing ratio in nearly all cases of pollution is close to unity. The pressures of the gases P_{NH_3}, P_{SO_2} and P_{CO_2} are the same as the molar mixing ratios $[NH_3]$, $[SO_2]$ and $[CO_2]$ when the pressures of these gases are in atmospheres; numerically the values are the same as molar mixing ratios at one atmosphere total pressure.

Solution of this equilibrium set is achieved by realising that the system must be electrically neutral, that is, positive charges must equal negative charges:

$$[NH_4^+] + [H^+] = [OH^-] + [HSO_3^-] + [HCO_3^-] + 2[SO_3^=] + 2[CO_3^=] + 2[SO_4^=], \tag{4.5}$$

where the $[SO_4^=]$ results from the oxidation process. We substitute for the various terms and get an equation of the form

$$a[H^+]^3 - b[SO_4^=]\,[H^+]^2 - c[H^+] - d = 0, \tag{4.6}$$

where *a, b, c* and *d* are constants. This cubic equation determines the hydrogen ion concentration, $[H^+]$, or the pH of the solution. The formation of $SO_4^=$ in solution occurs by oxidation of SO_3^- ion such that:

$$\frac{d[SO_4^=]}{dt} = k[SO_3^=] \tag{4.7}$$

with $k \approx 0.1\ \text{min}^{-1}$ at 25°C.

Final solution is a combination of an integration of this equation with the pH equation 4.6. At first, $[SO_4^=]$ is considered absent and set to zero so the $[H^+]$ is obtained from the cubic equation without the second term. Then $[SO_3^=]$ is calculated from the combined equilibrium expression:

$$[SO_3^=] = \frac{K_{1s}\ K_{2s}\ [SO_2]}{K_{hs}\ [H^+]^2}$$

and $d[SO_4^=]/dt$ calculated. Then, in the simplest, Euler approximation:

$$[SO_4^=] \approx \frac{d\,[SO_4^=]}{dt}\,\Delta t, \tag{4.8}$$

where Δt is the selected time step. With the $SO_4^=$ concentration calculated, $[H^+]$ is recalculated and the $[SO_3^=]$ concentration recalculated. The reaction (equation 4.7) results in more $SO_4^=$ which is added to the $SO_4^=$

already there and the [$SO_4^{=}$] gradually builds up with time.

The results of this calculation show the extreme dependence of the oxidation on pH, or the amount of NH_3. This creates a feedback for, as the $SO_4^{=}$ builds up, so does the H ion. This lowers the amount of $SO_3^{=}$, slowing the oxidation. The result is that the [$SO_4^{=}$] tends to approach a limiting value with time.

4.3 Nitrogen compounds

Particularly in smog episodes, nitrogen species come to the foreground. Five principle nitrogen-containing gases exist in the atmosphere:

N_2 – molecular nitrogen
NH_3 – ammonia
N_2O – nitrous oxide
NO – nitric oxide
NO_2 – nitrogen dioxide

and two species are formed in the condensed phase:

NH_4^+ – ammonia
NO_3^- – nitrate

Other oxides such as NO_3 and N_2O_3 are only important as intermediates in reactions though organic nitrates are sometimes formed in urban atmospheres.

Molecular nitrogen, N_2, is a permanent gas in the atmosphere yet some 100 tonnes a year are converted to reactive species in lightning strikes. Another apparently small amount of N_2 is converted to reactive nitrogen by nitrogen-fixing bacteria. N_2O is also a relatively inert gas though it is also produced and consumed by soil bacteria. It has a relatively constant concentration of about 0.25 ppm, and its main contribution to pollution is decomposition in the stratosphere, producing N_2 and O or NO and N. Generally, it may be excluded from the smog producers, but does influence the ozone balance in the upper stratosphere.

Ammonia, NH_3, is generally a useful though extremely variable nitrogen chemical. About 85 per cent of the NH_3 produced in the United Kingdom has been attributed to the NH_3 released from animal urine. It is a known product of the deamination of protein and in the oceans there are bacteria capable of producing NH_3 from nitrates. In its way it is the 'de-acid' agent of the atmosphere as it combines in the gaseous phase with H_2SO_4 and SO_2 and neutralises their acidity. As mentioned, it also tends to catalyse the oxidation of SO_2 in water droplets. One way or another it tends to form the ubiquitous ammonium sulphate particles that are major components of almost every aerosol; $(NH_4)_2SO_4$ is a good fertilizer.

About 3 per cent of the NH_3 is man made, being manufactured by the Haber process from N_2 and H_2; it is even injected directly into the soil as fertilizer. In the atmosphere NH_3 is quickly removed by clouds, rain or acid gases and so is highly variable. In the lower levels of the atmosphere it has a residence time of a few hours though in a clear, dry environment it may remain for weeks.

The nitrogen species, NO, NO_2, combined with hydrocarbons, produce ozone and undesirables that are eye irritants and pose health problems (Galloway *et al.*, 1985; Goldberg, 1982). Generally NO is formed during high temperature combustion, principally from motor vehicle exhausts but including stationary combustion (Eskinazi *et al.*, 1989). The NO reacts in the stewing pot with hydrocarbons and sunlight to produce NO_2 which in turn reacts with a host of reactive species to produce quite undesirable materials but principally ozone. We require:

(i) a lid or atmospheric inversion on the pot of chemicals;
(ii) high motor vehicle traffic (or coal burning); and
(iii) sunlight

to produce photochemical smog. In Sydney and Perth this requires 'a perfect day', clear skies, a light wind, an afternoon sea breeze (Carras and Johnson, 1983).

Though the exact chain of events is disrupted, the reactions that follow are well known (after Butcher and Charlson, 1972; Wayne, 1987);

(i) Nitric oxide formation

$$NO_2 + \text{UV light} \rightarrow NO + [O]$$
$$\phi k_1 = 25 \text{ hr}^{-1} \; (\lambda < 380 \text{ nm})$$

(ii) Ozone production

$$[O] + O_2 + M \rightarrow O_3 + M^{*}$$
$$k_2 = 9\; 10^{-4} \text{ ppm}^{-2} \text{ hr}^{-1}$$

(iii) Ozone removal

$$O_3 + NO \rightarrow NO_2 + O_2$$
$$k_3 = 1{,}320 \text{ ppm}^{-1} \text{ hr}^{-1}$$

(Note that [O] here means reactive, nascent oxygen atoms where the ks are for 25°C and 1 atmosphere. The term ϕ represents a maximum radiation level at 45° latitude at noon on the summer solstice.)

Figure 4.2 shows the amount of NO expected to be in a combustion chamber with 20 per cent fuel. Time is required for equilibrium to be attained with the effect that rapid heating produces little NO and rapid cooling does not allow for NO destruction. The worst case is slow heating

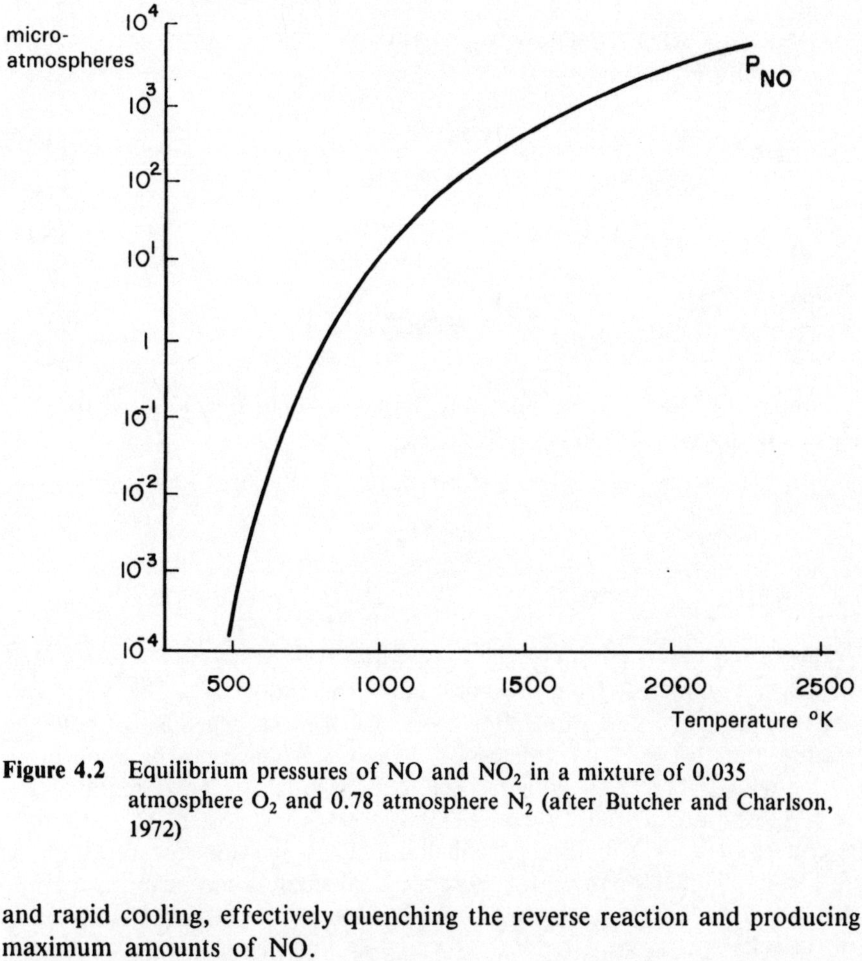

Figure 4.2 Equilibrium pressures of NO and NO_2 in a mixture of 0.035 atmosphere O_2 and 0.78 atmosphere N_2 (after Butcher and Charlson, 1972)

and rapid cooling, effectively quenching the reverse reaction and producing maximum amounts of NO.

Nitric oxide is man's greatest single contribution to the pollution but he also produces a myriad of other reactive organic radical species so that within a few hours most of the NO is converted to NO_2. The production of O_3 by the mechanism shown is relatively rapid, i.e. when the NO and O_3 concentrations are 0.05 ppm the reactions are complete in about 16 seconds. Of course, depending on the bits and pieces in the stew and the height of the sun in the sky, the reaction continues to produce O_3 through most of the day.

The net rate of accumulation of a species is the rate of input minus the rate of output. For each of the species present this mass balance can produce a differential equation. With the initial conditions, the formulation is complete. In case of photochemical smog, mass balances of NO_2, NO, O and O_3 lead to the following set of simultaneous differential equations:

$$\frac{d[NO_2]}{dt} = k_3[O_3]\,[NO] - k_1\phi[NO_2] \qquad (4.9)$$

$$\frac{d[NO]}{dt} = \phi k_1[NO_2] - k_3[NO]\,[O_3] \qquad (4.10)$$

$$\frac{d[O]}{dt} = \phi k_1[NO_2] - k_2[O]\,[O_2]\,[M] \qquad (4.11)$$

$$\frac{d[O_3]}{dt} = k_2[O]\,[O_2]\,[M] - k_3[NO]\,[O_3] \qquad (4.12)$$

Assuming ϕk_1 and k_3 are relatively large, a quasi-steady state for NO obtains such that $d[NO]/dt = 0$. Then

$$\phi k_1[NO_2] = k_3[NO]\,[O_3]$$

$$[O_3] = \frac{\phi k_1[NO_2]}{k_3[NO]}. \qquad (4.13)$$

This is the photostationary state expression for ozone and suggests that the concentration of ozone depends upon the ratio $[NO_2]/[NO]$ for any value of $\phi k_1/k_3$. The maximum value of ϕk_1 is dependent upon the latitude, time of year and time of day. Table 4.1 illustrates the importance of the ratio $[NO_2]/[NO]$ with respect to how much ozone is required for the photostationary state to exist. This suggests that most of the NO must be converted to NO_2 before O_3 will build up in the atmosphere (Stern *et al.*, 1984). The daily pattern of photochemical smog is shown in Figure 4.3. Here we have a rise in the NO concentration accompanying the morning traffic followed by oxidation of NO to NO_2, which, in turn, is followed by an increase in O_3 and the appearance of other nasty substances, the organic nitrates, oxygenated hydrocarbons and carbon monoxide, CO.

It is not really known what goes on in the stewpot but certainly assorted forms of elemental oxygen and free radicals such as HO_2 and OH react with hydrocarbons and produce intermediates. It is also possible that SO_2 is involved.

However, the principal nitrogen-containing organic compounds produced are the peroxy-nitrates, one of which is PAN, peroxyacetal nitrate:

$$CH_3 - \overset{\overset{\large O}{\|}}{C} - OO - NO_2$$

Table 4.1 $[O_3]$ predicted from photostationary state approximation

Initial $[NO_2]$ (ppm)	Final $[NO_2]$ (ppm)	Final $[O_3]$ (ppm)	$[NO_2]/[NO]$
0.1	0.064	0.036	1.78
0.2	0.145	0.055	2.64
0.3	0.231	0.069	3.35
0.4	0.319	0.081	3.94
0.5	0.408	0.092	4.43
$\phi k_1 = 0.5\ \text{min}^{-1}$, $k_3 = 24.2\ \text{ppm}^{-1}\ \text{min}^{-1}$			

Source: after Stern *et al.*, 1984

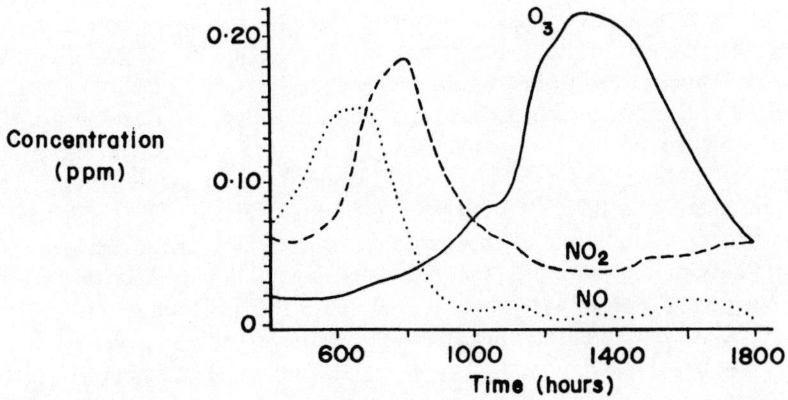

Figure 4.3 Average daily patterns for the time dependence of NO, NO_2 and O_3 (after Butcher and Charlson, 1972)

which is highly reactive and dangerous. It reacts, in turn, with NO in the dark or in the condensed phase with basic compounds, OH^-.

The final step in its production is thought to be

$$CH_3 - \overset{\overset{\displaystyle O}{\|}}{C} - OO{:} + NO_2 \rightarrow PAN$$

peroxide radical

The noxious elements are many but include principally:

(i) NO_2
reddish brown – worst irritant of nitrogen oxides – which forms HNO_3 with water:

$$3NO_2 + H_2O = 2\ HNO_3 + NO$$

An NO_2 concentration of 10 ppm reduces visibility to 1.6 kilometres. Concentrations of 0.2 ppm start oxidation of organic material.

(ii) O_3, ozone. Maximum allowable WHO concentration is 0.06 ppm. Sydney has experienced concentrations as high as 0.4 ppm; Los Angeles, 0.7 ppm.

(iii) PAN as well as other organics including relatively noxious aldehydes and ketones, some of which are directly produced by combustion.

(iv) Minute particles produced by poor burning or by gas-to-particle conversion. They are responsible for the visible grey layer.

The result is that people experience eye irritations that may be so severe that driving is no longer possible. Plants suffer and become burnt, and crop yields are reduced; in particular, tomatoes and petunias are affected. The paint on homes deteriorates, as do rubber products.

Note that the one product, ozone, is naturally produced and consumed in the presence of certain wavelengths of ultraviolet light. In the upper stratosphere, above about 20 kilometres, where UV (ultraviolet) light levels are high, there is a layer of high concentration (Isaksen, 1988). This layer is thought to act as a shield that removes harmful UV from the incoming solar radiation and currently there is some alarm that this layer may be being depleted, possibly due to the presence of man-made chlorinated hydrocarbons; these compounds are stable in the lower atmosphere but are broken up by UV light when they reach the upper stratosphere, decades after their release. The active chlorine atoms and radicals consume the ozone and lower the ozone concentration levels.

The result is likely to be an increase in UV light at the surface of the earth, affecting human bodies and the environment. In particular, the added UV light is expected to produce more skin cancers, cataracts and a general lowering of the ability of the immune system to combat disease. Natural ozone itself is not limited to the stratosphere. On special meteorological occasions, stratospheric air may intrude into the troposphere and bring the ground levels of ozone to 30 ppm.

4.4 Carbon compounds

(i) Carbon dioxide

In one form or another CO_2 is represented in the atmosphere, hydrosphere and lithosphere. In primeval times it is thought that the atmosphere itself was primarily CO_2. Since those early times most of the CO_2 has been trapped in the lithosphere as carbonates. At this time the carbon distribution is roughly

1 part in the atmosphere as CO_2
30,000 parts in the lithosphere as CO_3^-
10,000 parts in the lithosphere as reduced carbon (coal)
60 parts in the hydrosphere as ions in the deep sea
1.2 parts in surface waters
0.5 parts as land plants
1.7 parts as humus.

The carbon dioxide levels in the atmosphere are increasing at a rate greater than 1 ppm per year. Present levels are about 350 ppm. Since CO_2 and other 'greenhouse' gases such as CH_4, H_2O absorb IR (infra-red) radiation, increases in its concentration may cause a general warming of the earth (Gregory, 1988; Pearman, 1989; Idso, 1989; Hileman, 1989). The net effect may be an increase of 0.5°C in the mean temperature of the earth by the year 2000. However, though the CO_2 levels are increasing, the mean temperatures actually fell between 1940 and 1970. This may be an effect of blanketing due to aerosols or, as has been suggested, increased absorption of the IR radiation by the oceans, higher water vapour in the atmosphere with increased deposition of snow on the poles and, thence, cooling due to the increased polar reflectivity. Nature is not simple and CO_2 is an important constituent. Today it is believed the mean temperatures are increasing generally due to increases in CO_2 levels.

It has been proved that the main source of the increase is anthropogenic because the ratio of carbon-12 to carbon-14 increases with time. This ratio is found by analysing the carbon isotopic ratios in tree rings. Carbon-14 is produced by cosmic ray bombardment of air molecules and is trapped in the tree rings during growth. Fossil fuels, of course, have resided in the ground for long time-periods and the radioactive carbon-14 has diminished, leaving a significantly larger amount of carbon-12 in exhaust emissions from fossil-fuel burning.

Other natural phenomena that may affect CO_2 concentrations are: (i) a decrease in the CO_2 level of surface water, (ii) an increase in the oxidation of plants (deforestation), and (iii) increased photosynthesis of plants. Environmental changes of CO_2 levels are a result of many complex interactive phenomena.

(ii) Carbon monoxide

This is the one pollutant which is mostly produced by mankind, who generates about 70 per cent of the total. It is remarkably stable in the atmosphere, being consumed only in the lower stratosphere, by biological removal in soil and a few photochemical reactions.

In a city almost 100 per cent of CO is anthropogenic, the CO being produced by fossil-fuel consumption and motor vehicles (hence it is almost totally carbon-12). Background levels of CO are ~ 0.1 ppm; urban levels ~ 10 ppm. As with most carbon compounds, motor vehicles account for about 60 per cent of the total. It is noteworthy that stationary sources, power plants, gas stoves, etc. contribute little CO to the atmosphere. This is probably because the burning efficiency is much greater. CO is produced when the fuel to oxygen ratio is too high or the temperature of combustion is too low, slowing down the oxidation process. Motor vehicles are normally operated at a slight excess of petrol, which gives maximum power and performance, and produces high CO levels.

CO concentrations are highly variable in cities, with maxima occurring at the peak traffic times. When measurements are made immediately downwind of a street, it is possible to observe plumes from individual vehicles.

When inhaled, CO enters the system as if it were oxygen; it has about 200 times the affinity for combining with haemoglobin as does oxygen. It combines with haemoglobin to form carboxy haemoglobin, greatly reduces the capacity of the blood to carry oxygen and, in acute cases, asphyxiation results. Long-term exposure to 100 ppm causes behavioural changes, decreased visual performance, cardiovascular effects and general euphoria.

The main reaction of CO in the atmosphere is probably with hydroxyl radical:

$$CO + [OH] \rightarrow CO_2 + [H].$$

The hydrogen can participate in a chain of reactions, details of which are given by Demerjian *et al.* (1974).

4.5 Organic compounds

(i) Methane

A most abundant organic compound, with a natural level between 1 and 2 ppm that is everywhere being produced during the rotting of dead plants in swamps and natural and artificial leaks from natural gas pockets. Yet it is relatively unreactive. Sinks for methane are unknown but probably are biological.

(ii) Terpenes

Emitted from plants as a part of plant (tree) life processes, particularly when they are damaged or defoliated. They include β-pinene, β-pinene, myrcene, limonene; these are monocyclic and dicyclic organic compounds built up from isopene

$$\begin{array}{l} CH_2 = \underset{\displaystyle CH_3}{\underset{|}{C}} - CH = CH_2. \end{array}$$

The terpene α-pinene is made up from two of these groups (ten carbon atoms); it is most abundant in the northern hemisphere. Since these compounds contain one or more double bonds they are extremely reactive with photochemical species, including ozone. They are, in fact, a source of haze in scarcely populated areas or urban areas that are thickly wooded. They may be a major sink for O_3; their reaction rate is such that their concentrations are around 50 $\mu g\ m^{-3}$ even though their emission rate is comparable to that of CH_4.

(iii) Other natural organics

One cannot ignore the stench of rotten eggs from the methyl and dimethyl mercaptans (CH_3SH, $(CH_3)_2S$) emitted from rotting biota (seaweed) at the seashore. In addition methyl iodide (CH_3I) is emitted from seaweed and, in fact, is used as a commercial source of iodine. Also, when combined with lead to produce lead iodine it forms a most effective ice nucleating agent. It is by this route that lead from motor vehicles may influence cloud properties and thus climate.

Other organics of note are bacteria, fungi, pollen and spores, as well as soot and burned material from bush and forest fires.

(iv) Anthropogenic hydrocarbons

Man generally releases about 10 per cent of the non-methane hydrocarbons found in the atmosphere. Approximately half of these emissions are from motor vehicles; 10 per cent, solvent evaporations; and at least 10 per cent from intentional bush fires or stubble-burning. Thus the concentrations of these organic substances follow the same diurnal trends as CO. From these hydrocarbons a multitude of secondary products are created in smog conditions. These include formaldehyde

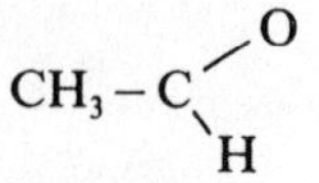

and acrolin

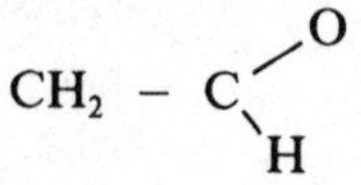

and other doubly-bonded organics that are highly reactive themselves. In the home and office, formaldehyde foam insulation and urea-formaldehyde glues are used in chipboard and plywood. Owing to the energy crisis there is a tendency to seal homes, trapping in these pollutants; measured values of total hydrocarbons have been found to exceed maximum recommended levels in homes which have only one air change in five hours.

Of course ash and carbonaceous material is a common emission of incinerators and fires of all sorts, as well as being a component of smog. Since different sources emit different proportions of different hydrocarbons (and other pollutants), measurements of several different hydrocarbons can actually identify the source of the material.

Other compounds of concern are those that are long-lived in the atmosphere. These include the modern pesticides and herbicides; metal organic compounds, dimethyl mercury (industrial) and tetraethyl lead (motor cars), as well as the freons, chlorofluorocarbons, a group of fully chlorinated and fluoridated hydrocarbons that are remarkably inert (Freon 12 is CCL_2F_2). These compounds are used as refrigerants and propellants in aerosol cans. They are worrying in that they survive long enough to enter the stratosphere where they release chlorine compounds that tend to consume ozone from the protective ozone layer.

4.6 Aerosols

An aerosol is a gas containing particles with a small settling velocity. Thus the atmosphere is an aerosol since it always has some particulate matter of small size in it (Hidy, 1984). In a smoke-filled room there may be 1,000,000 to 10,000,000 particles per cm^3! In an ordinary city the particulate concentration would be 10,000 to 50,000 cm^{-3}. In a clean location, remote from sources, concentrations as low as 100 and even 20 cm^{-3} can be obtained.

A size distribution of aerosols is presented in Figure 4.4. On the figure's vertical axis is the number of particles per cm^3 of air per logarithmic size interval. This is not easy to understand but is the equivalent of a number histogram produced by measuring sizes of particles collected from a cm^3 of air using sieves, each of which has a size an order of magnitude

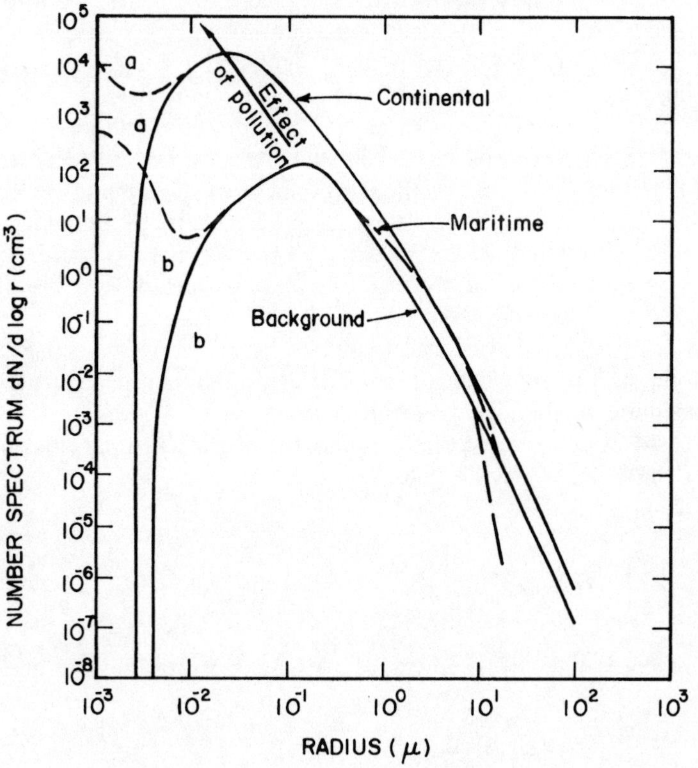

Figure 4.4 Size distribution of aerosols (from Ruskin and Scott, 1974)

smaller than the sieve above. The histogram is simply smoothed to produce the continuous distribution shown.

The figure shows several interesting features of the atmospheric aerosol. Firstly, it tends to be logarithmic and most easily plotted on a log–log graph. Secondly, most of the particles are in the smaller particle sizes. As shown, a continental air mass has larger numbers of smaller particles; a marine air mass has few. Pollution tends to add the smallest particles by gas-to-particle conversion processes (a). Coagulation tends to create larger particles from these smaller particles (b). Specifically, particles in the atmosphere generally range from 10^{-2} μm to 10^{2} μm. Particles show a maximum of number in the size range around 0.1 μm; a maximum in mass around 1 μm. Particles of size below 0.1 μm are called Aitken particles after John Aitken, who developed the Aitken counter in the early 1900s; they are also termed 'small particles'. Particles above this size are considered 'large particles' though the term 'giant particles' is used for particles above 1 μm and 'gigantic particles' for particles above 10 μm.

The smaller particles that are hydroscopic are effective in warm cloud development and are called cloud condensation nuclei; particles ~ 0.5 μm are of the order of the wavelength of light and are optically effective in scattering processes. The gigantic particles have an appreciable fall velocity (0.1 ms^{-1} at 40 μm diameter) and are 'precipitation embryos'.

Instead of size information, a similar diagram could be made regarding chemical composition, that is, the fractional mass of sulphate ion contained in a given size plotted versus size. Or, perhaps, particles of a given shape, or, say, the fraction that are pollen grains, etc. might be used.

A few characteristics of the size distribution function are:

(i) Mode – peak value of distribution

(ii) Mode radius – value of particle radius that corresponds to the maximum in the size distribution function.

(iii) Median radius – value of the radius for which half the values lie above and half the values lie below.

That is if:

$$N_T = \int_{r=0}^{\infty} f(r)\, dr$$

then N_T is the total number of particles per unit volume and

$$\frac{1}{N_T} \int_{r=0}^{r_m} f(r)\, dr = 0.5$$

where r_m is the median radius.

(iv) Number mean radius; the first moment of distribution

$$\overline{R} = \int_0^{\infty} r f(r)\, dr$$

(v) Volume mean radius, the third moment of distribution

$$R_{vm} = \sqrt[3]{\int_0^{\infty} r^3 f(r)\, dr}$$

This last quantity is usually assumed to be the mass mean diameter, which presumes that the density of all particles is the same.

It is useful to be able to describe size distribution by a simple mathematical expression. The simplest description is a two-parameter function in the form of a power law. One of these in common use is the Junge Size Distribution, suggested by C. E. Junge in the late 1950s. He found that atmospheric aerosol particles of sizes between a tenth of a micron to a few tens of microns often have a size distribution with a constant volume per log radius interval. That is, he found that the total volume of particulate material between 0.4 and 0.6 μm, 0.6 μm and 0.9 μm, 1 and 1.5 μm and 4 and 6 μm was approximately the same despite the 1,000,000 fold change in the volume of a single particle in this size range. This parametric form may be written as

$$\frac{4}{3}\pi r^3 \frac{\Delta N}{\Delta \log r} \text{ or } \frac{4}{3}\pi r^3 \frac{dN}{d(\log r)} = \text{constant.}$$

Or, in natural logarithms,

$$d\ln r = \frac{dr}{r},$$

$$\text{which means that } \frac{dN}{dr} \propto r^{-4}.$$

This general approach (allowing that the parameters may vary) suggests a density function of the form

$$f(\log r) = \frac{dN}{d(\log r)} = Cr^{-\beta},$$

where β varies between 2 and 3 normally.

Aerosol particles are produced by a number of processes, including (i) gas-to-particle conversion, (ii) growth and coagulation, (iii) breaking waves on the ocean, (iv) wind-blown (Aeolean) effects, (v) direct emissions, and (vi) meteorites. Scanning the size distribution curve (see Figure 4.4), the smallest particles are produced by gas-to-particle conversion processes (a), the effect of pollution being to produce more reactive vapours that, in turn, form more particles. If the vapour concentrations are sufficiently high, particles may be formed in the free air by 'homogeneous nucleation'. Otherwise, if there are sufficient particles present, the vapours condense on the particles.

In any case, the particles formed are generally small. They grow by further incorporation of vapours and, more importantly, incorporation of smaller particles, which 'coagulate' to produce particles 0.01 to 0.1 μm (b). Above a size of 1 μm directly emitted material from motor cars, industry, volcanoes and breaking waves becomes important. There is no single important source of particles between 0.1 and 1 μm and in this region removal

Table 4.2 Aerosol properties

Radius	Fall velocity (quartz) cm s^{-1}	Brownian diffusion coefficient D cm^{-2} s^{-1}	Residence* time τ	Removal process
0.001 μm	-	10^{-2}	~ 1 day	coagulation
0.01 μm	4×10^{-7}	10^{-4}	~ 1 week	coagulation
0.1 μm	0.00015	2×10^{-6}	-	nucleation
1.0 μm	0.025	10^{-7}	~ 1 mo.	sedimentation
10.0 μm	2.5	-	½ day	sedimentation
100.0 μm	150	-	10 min.	sedimentation

*The residence time has been estimated for the most important removal process, a 1 kilometre fall distance was used.

Source: estimates from Twomey, 1977

mechanisms are so ineffective few need to be produced to generate the large concentrations observed (see Table 4.2).

The breaking of waves on the open ocean is a major source of particles between 0.5 and 5 μm. In this instance bubbles become trapped in the water during the breaking of waves. These bubbles on emerging from the surface form a thin, liquid film on their upper surface, which breaks in a crown-like formation, ejecting hundreds of droplets into the air. On drying, particles of about 0.5 μm are formed. The cavity left by the broken bubble is then filled by an inrush of water, which collides in the middle to form a Rayleigh jet, which breaks up into larger droplets, which in turn produce larger particles. All these salt particles are extremely important in cloud processes and are thought to ultimately form the embryos of precipitation.

The higher level of aerosols in continental air can be attributed to larger natural and man-produced pollution levels including aeolean (wind) effects, emissions from burning, and gas-to-particle conversion processes. The largest particles (10 μm or greater) have reasonable fall velocities and short lifetimes and so are fleeting components of the atmospheric aerosol. The single most common source of the largest particles is meteor showers.

Of particular interest is the production of wind-blown dust. It is surprisingly difficult to dislodge particles from a surface. In fact, the classical work in the matter is a study by Bagnold (1941) and recent work has been reviewed by Nicholson (1988a). He found that the air blown over finely divided cement could not dislodge particles though pebbles 4.6 mm in diameter would move from the wind force.

To treat the matter, we return to the atmospheric boundary layer, the viscous sublayer and a grain, sitting on the surface. To lift it, we must have a force equal to:

$$F = \varrho_p \frac{\pi}{6} d^3 g, \tag{4.14}$$

where ϱ_p is the density of the grain of diameter d. This force acts through a moment arm $\approx d$, so the moment required to twist the particle from the surface is

$$M \; \alpha \; Fd \; \alpha \; d^4 \varrho_p g. \tag{4.15}$$

Similarly, the aerodynamic force may be approximated by

$$\tau_0 \pi \frac{d^2}{4},$$

and the aerodynamic moment is approximately $\tau_0 \pi d^2(d)/4$ where τ_0 is the value of the horizontal shear stress at the surface. Equating the aerodynamic moment to the moment required for movement, it is found that the required shear stress is

$$\tau_0 \approx 0.01 \; d \; \varrho_p \; g,$$

where the constant is experimentally derived.

Now the drag coefficient C_d may be defined by $\tau_0 = C_D \varrho u^2$; the drag coefficient varies from a small value (0.001) for smooth terrain to a value approaching unity in rough terrain.

Substituting, the final equation for the mean horizontal wind (at 2 m) required to lift a particle of size d is

$$u \approx \sqrt{\frac{\varrho_p d}{C_D}} \tag{4.16}$$

where SI units are used. In using this expression, one must understand that it contains a mixture of time and space scales and that only eddies of size comparable to the size of the particles will be able to lift them, so that apart from the rough analysis given, there must be sufficient energy in scales of the size of the particles.

These small-scale eddies dissipate rapidly and, of course, may be produced by the very elements they may dislodge. The result is that particles of small size tend to be associated with very small values of C_D and require very large wind speeds to become airborne; particles of large size, however, are associated with large values of C_D and so require only moderate winds to become airborne. The effective C_D values approach a limit with the largest particles, however, so that the numerator term d becomes important (their weight is too great) and they cannot be lifted. Compounding the difficulty of moving the smallest particles is the fact that the flow field effectively becomes non-turbulent and the drag force is then the Stokes form which is linearly related to particle size.

Adhesion also plays an important role and tends to be greater in a static situation than in a dynamic state. The actual release of particles probably occurs through 'saltation', whereby particles roll over each other with a jumping, bouncing action. The effect reduces the adhesion force, which tends to be roughly proportional to particle size, and helps to inject particles outward into a zone (Buffer layer) where turbulent eddies may carry them into the main flow. Of course, the viscous sublayer decreases as the wind speed increases and, in the case considered, one would expect the shear near the surface to roll particles of size around 40 μm and that these particles would be easily dislodged.

In fact, it is difficult to make particles less than about 40 μm airborne. Smaller particles must be produced during saltation by violent disruptions that may even fragment larger particles or part of the surface. Even if this effect is taken to an extreme in mechanical crushing, however, it is not possible to produce particles of size much smaller than a micron. Adhesive and cohesive forces are too great.

The distance a particle can move upwards when injected vertically at a velocity w_0 is limited by viscous drag, governed, more or less, by an equation of the form

$$m \frac{\mathrm{d}V}{\mathrm{d}t} = 6\pi a\eta\,(v_f - v) + m\,g, \tag{4.17}$$

where v_f is the (vertical) velocity of the fluid which is insignificant near the ground, mg is the force due to gravity, a is the particle radius, m is the particle mass, and v the particle velocity. (Note that a similar equation applies to the horizontal velocity – without the term mg, of course.)

Substituting $\mathrm{d}t = \mathrm{d}z/v$ and separating variables,

$$\mathrm{d}z = \frac{v\mathrm{d}v}{\dfrac{6\pi a\eta v}{m} + g}.$$

Integrating from initial velocity w_0 at $z = 0$ to final velocity $v = 0$ at $z = d_s$, we get

$$d_s = \frac{1}{A}\left[w_0 - \frac{g}{A}\ln\frac{(Aw_0 + g)}{g}\right], \tag{4.18}$$

where $A = 6\pi a\eta/m$.

This is the vertical stop distance which gives the vertical distance a particle will go in a fluid if its initial velocity is w_0. The horizontal stop distance can be derived in zero horizontal wind; the equation for the stop distance is the same, it simply does not contain the gravity term. In the case of a horizontal wind, a numerical calculation of the trajectory is necessary.

After injection into the outer, turbulent airstream, turbulent eddies also move the particles. Considering only the vertical dimension, the mass balance for the particle concentration also contains a divergence term for eddy diffusion, which tends to convey particles upward from the surface. The net mass balance for particles contains the downward, opposing influence of fallout and leads to the time dependent partial differential equation.

$$\frac{\partial n}{\partial t} = K \frac{\partial^2 n}{\partial z^2} + V_g \frac{\partial n}{\partial z}, \tag{4.19}$$

where V_g is the terminal velocity (fall speed in steady state), n is the particle concentration, and K, the eddy diffusion coefficient.

This equation presumes that 'K-theory' can be applied to the conveyance of particles; that is, particles are moved by the turbulent motions in a manner that transports them down the concentration gradient similar to the molecular diffusion of molecules. With very small particles, this is a good approximation and, it appears that when Brownian motions are important (below about 0.1 μm) the analogy is exact. Practically all particles of size below 10 μm react quickly enough (have small reaction times and stop distances) to be considered as being conveyed by this mechanism. Of course, the K values may not be the same as the values for momentum, especially if the particles are of such a size that they produce significant eddy motions themselves, from differential motions and wake effects.

In the steady state, without the time-dependent term, this equation has a simple solution that is well known in sedimentology. The equation is integrated with $n = n_0$ near the surface ($z = z_0$) and the condition of zero eddy flux at great heights. For neutral conditions when $K = ku_* z$, the solution is

$$n = n_0 \left(\frac{z}{z_0}\right)^{-\gamma}, \tag{4.20}$$

Where γ is the atmospheric equivalent of the Rouse number of sedimentology, $\gamma = V_g/ku_*$. Surprisingly, this fits concentration profiles well, especially in the case of appreciable wind erosion. Usually $\gamma \approx 1$ suggests that there is a feedback or winnowing effect that removes smaller particles from the surface with time. Note that the profile depends on the relative

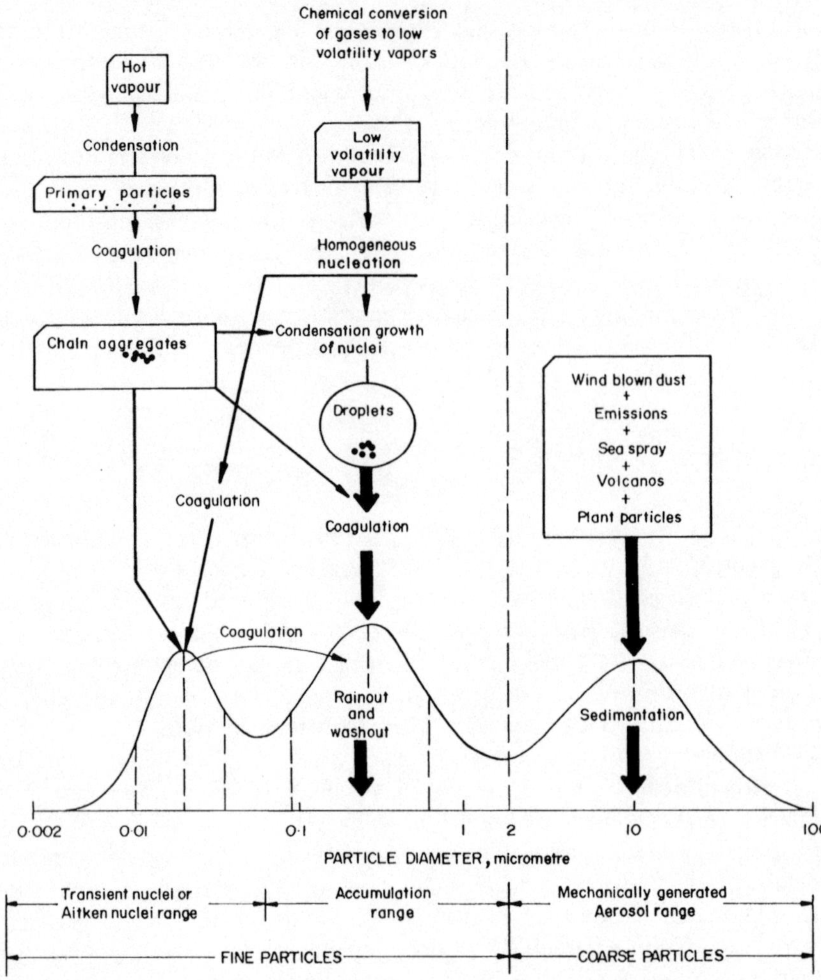

Figure 4.5 A postulated atmospheric aerosol surface-area distribution (after Whitby, 1975; Drake *et al.*, 1979). Copyright © 1979 Electric Power Research Institute. Reprinted with permission.

effect of particle size (through V_g) and surface shear (through u_*). Hence 'fluffy' particles can be moved in a light breeze, whereas heavy particles require a gale. Also note that the scale of the profile is set by the roughness, z_0; and that large particles, with large values of γ have concentration profiles that drop to nearly zero close to the surface.

Particles are involved in a multitude of processes in the atmosphere (Butcher and Charlson, 1972). A few of the important removal processes are presented in Figure 4.5. Note that this is an area-based plot of interactions – because most of the processes in the smaller size range are determined by the surface area of the particles. The three peaks in the distribution result from a combination of source effects, transportation both by advection, and through the size distribution by coagulation, capture, etc., and, ultimately, removal. The principal sources of fine particles are gas-to-particle conversion. The bimodality of this size range results from the great difficulty of getting particles of sizes around 0.1 μm together; it is nearly impossible to manufacture such particles mechanically.

Gases in the atmosphere are transparent, aerosols are not. Hence every air-pollution problem as seen by the public eye, every observation of 'dirty air', is due to the presence of aerosol particles. This does not mean that they are, indeed, the harmful agents of the pollution, just that their presence tends to be aesthetically displeasing. An example is a diesel-powered vehicle, pouring out black smoke. The particles are in the size range 10–100 μm, fall out rapidly, and few enter the human respiratory system. Gases and very small particles, however, are nearly invisible yet they penetrate deeply into the lungs.

With aerosols, one must also realise that only particles that have sizes comparable to the wavelength contribute significantly to optical effects. The smaller particles scatter similarly to gases, the larger particles simply act to extinguish the light by mechanical obstruction, effectively producing shadows.

4.7 Kinetic modelling

Until now most of the processes have been considered in isolation. The last figure exemplifies the real situation with many chemical and physical influences. A full mass balance requires that *all* processes be considered and we usually hope that at least the important ones are considered. Even in the use of the chemical kinetics, however, the equations of mass balance are not simple.

Our knowledge is usually insufficient to quantify most of the reactions, hence various workers have attempted to model the reactive atmosphere through rate equations and adjusted the rate parameters in line with observations. For example, Friedlander and Seinfeld (1969) assumed that the dynamics of smog formation could be approximated by a series of simple reactions. Unfortunately, they were unable to account for various classes of organic compounds and their treatment of free radicals was grossly oversimplified. Subsequent authors (Hecht and Seinfeld, 1972; Hecht *et al.*, 1974; Eschenroeder and Martinez, 1972) have addressed these problems but the models invariably introduce additional empirical parameters.

Consider the simple set of chemical reactions

$$A + B \xrightarrow{k_1} C$$

$$C + D \xrightarrow{k_2} A$$

Ignoring any spatial dependence, each of the mass balances for A, B, C and D gives, respectively, the following set of simultaneous, differential equations:

$$\frac{d[A]}{dt} = -k_1 [A] [B] + k_2 [C] [D] \tag{4.21}$$

$$\frac{d[B]}{dt} = -k_1 [A] [B] \tag{4.22}$$

$$\frac{d[C]}{dt} = k_1 [A] [B] - k_2 [C] [D] \tag{4.23}$$

$$\frac{d[D]}{dt} = -k_2 [C] [D] \tag{4.24}$$

where the rate constants k_1, k_2 are determined either from separate experiments or empirically so as to fit the observations.

A general discussion of the fate of air pollutants cannot assume that all pollutants remain in the gaseous state as sometimes, gas-to-particle conversion removes gaseous constituents (for instance, oxidation of sulphur dioxide to sulphate). This removal process, or the process of selective capture by a surface 'sink' can be considered in the simplest formulation as first order kinetic reaction of the form,

$$A \xrightarrow{k} B$$

$$A \xrightarrow{k_a} C \text{ (surface)}$$

$$B \xrightarrow{k_b} C \text{ (surface).}$$

The net effect describes the removal process in its simplest form, by the equivalent equations of chemical kinetics

$$\frac{d[A]}{dt} = -k[A] - k_a[A] \tag{4.25}$$

$$\frac{d[B]}{dt} = k[A] - k_b[B] \tag{4.26}$$

where k is the reaction rate and k_a, k_b are the surface removal rates. Note that both the vaporous conversion of A and the surface conversion of A cause a net decrease in the concentration of A whereas B is affected by all these reactions. With initial conditions $[A] = A_0$ at $t = 0$, $[B] = 0$ at $t = 0$, the solution to these equations is

$$\frac{[A]}{A_0} = \exp\,(-\,(k + k_a)\,t) \tag{4.27}$$

$$\frac{[B]}{A_0} = \frac{k}{k + k_a - k_b}\,\{\exp\,(-\,k_b t) - \exp\,(-\,(k + k_a)\,t) \tag{4.28}$$

These equations may be appropriate to the oxidation of $SO_2(A)$ in droplets, where k may be 170 min^{-1}. If the removal rate of $SO_4(B)$ corresponds to the time it takes for rain to form, k_b is around 3 hr^{-1}. The removal of SO_2 by dry deposition is small but, if an appreciable amount of SO_2 is dissolved in the droplets, k_a may also be about 3 hr^{-1}. It is doubtful that the k values can be known with a sufficient accuracy but, in this example, the k_a and k_b effect cancels out in equation 4.28, making the $SO_4^{=}$ concentration mostly dependent on the reaction rate in the droplets.

The case of higher-order, multiple reactions is equally as deceptive. Consider the photochemical smog reaction:

$$O_2 + 2NO \rightleftarrows 2NO_2.$$

It can be broken up into the equivalent forward and reverse reaction set:

$$O_2 + 2NO \xrightarrow{k_f} 2NO_2$$

$$2NO \xrightarrow{k_r} O_2 + 2NO.$$

Consider carefully what we write when we use a chemical equation; generally a rate constant times a concentration (or concentration product) gives a 'number of reaction events per unit time per unit volume'. In this case, the forward rate of conversion requires that this quantity be multiplied by 2 to obtain the number of conversions of NO_2 molecules (or moles of NO_2) per second per litre. The reverse reaction, similarly, removes two NO molecules with each reaction event. This means that the net mass balance equation for NO must be written as

$$\frac{d[NO]}{dt} = -2k_f[O_2]\ [NO]^2 + 2k_r[NO_2]^2.$$

Note the squared concentration that indicates the probability of finding an NO molecule and the equally, necessary factor of 2 that must be there to treat the whole mass balance – *this is not an error*. Similarly, mass balance for O_2 may be written;

$$\frac{d[O_2]}{dt} = -\ k_f[O_2]\ [NO]^2 + k_r[NO_2]^2.$$

This, of course, is a trivial example because we are not usually interested in the concentration changes of O_2. It is present in such concentrations that the few molecules involved in these reactions will have no detectable effect.

Chapter 5

ENVIRONMENTAL MONITORING AND IMPACT

Our analysis of atmospheric processes has enabled us to suggest how pollutants are moved, dispersed and react in the atmosphere. The discussion has concentrated on industrial pollutants but similar techniques may be used to study the dispersion of herbicides, seeds, pollens and even some aspects of the migration of flying insects. For example, Pasquill (1974) reports that secondary outbreaks of foot-and-mouth disease have been found downwind of a primary outbreak in a zone fairly accurately defined by the range of surface-wind directions over a period of several days following the initial outbreak. These observations also showed that the secondary outbreaks veered slightly from the surface-wind direction, keeping to some degree within the expected effects of vertical spread and the influence of the Ekman spiral.

Swarms of locusts travel directly downwind with the vectorial mean wind throughout the vertical extent of the swarm, with the speed of travel up to that of the wind for large swarms (Pasquill, 1974). Examples of 'stratiform' or 'cumuliform' swarms have been observed with the latter only in conditions of vigorous turbulence. In contrast to the obvious control exerted by the vertical component of the air motion on the vertical spread of locusts, the swarms tend to maintain horizontal cohesion and counter the dispersive action of the atmosphere in the horizontal.

The application of the Gaussian or any other model to these situations as well as industrial sources is an academic exercise unless the model can be validated. That is, we need to compare the observed concentrations with those predicted, bearing in mind the limitations of the model in terms of its intrinsic assumptions, accuracy of parameterization and the overall accuracy of our observed concentrations. The accuracy of dispersion models has been reviewed by Hanna *et al.* (1978), Smith (1984), Benarie (1987),

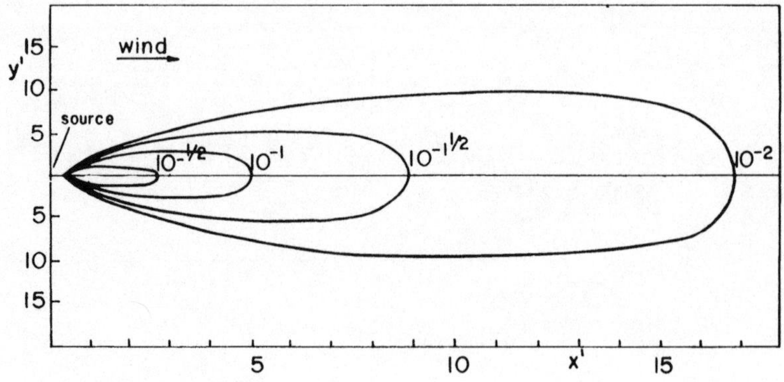

Figure 5.1 Idealized relative concentration isopleth from an elevated source for normalized along wind distance x′ and normalized across wind distance y′ for a constant relative plume spread of 2 (see Pasquill, 1974)

Table 5.1 Sensitivity of predicted concentration to input variables, where all values are for position at which change is maximum

Conditions resulting in modification of calculated χ	Changes in σ_z/h	
	+20%	−20%
(i) change in σ_z	+75%	−73%
(ii) as in (i) with 10° change in wind direction	−12%	−87%
(iii) change in h alone with 10° change in wind direction	+27%	−91%

Source: after Pasquill, 1974.

Hanna (1988, 1989) and Venkatram (1988), whereas Legros and Berger (1978) and Weber (1976) have carried out detailed sensitivity analyses on the Gaussian model with changes in input variables mathematically related to changes in predicted concentration.

The importance of a sensitivity analysis in specifying the desired input data accuracy can be seen by considering the concentration distribution from a single elevated source (Figure 5.1). This distribution is strikingly peaked and variations in concentration arise from changes in wind direction, effective height of the source and the rate of vertical spread. The resulting variations in the concentration at a fixed position relative to the source may be considerable. Thus concentrations at a fixed point are highly variable due to the stochastic nature of turbulence and the predicted concentrations from a model depend on the accuracy of the input parameters such as wind speed and wind direction.

An indication of the magnitude of concentration changes is shown in Table 5.1 for changes in selected input variables. Changes arising in this manner lead to a highly irregular record at a receptor located near the

Table 5.2 The scales of atmospheric motion

Motion	Rossby no.	Typical space scale	Typical time scale	Network spacing
Microscale	–	1 m	sec–min	cm–m
Mesoscale	10^2–10^1	5–10 km	hour	5–10 km
Small synoptic	1	100 km	hour–day	100 km
Large synoptic	10^{-1}	100–1000 km	days	100–500 km
Planetary	$< 10^{-1}$	> 1000 km	weeks	500 km

source. Nevertheless, model validation must relate concentration measurements to those predicted, at least to show the limits of validity of the monitoring. The underlying physics and chemistry should also be validated with and without the model, rather than the empirical constants within the model simply being adjusted to fit the observations. The techniques for evaluating the performance of air quality models have been reviewed by Bencala and Seinfeld (1979), KAMS (1982) and Keith (1987); here we will concentrate on the data input through monitoring.

5.1 Network design

The scale of the physical parameters being investigated have a gross effect on the network design. In the case of air pollution and concurrent atmospheric motion, a convenient scaling parameter is the Rossby number, Ro

$$\mathrm{Ro} = \frac{V}{fL} \tag{5.1}$$

where V is a characteristic velocity, L is a characteristic length and f is the Coriolis parameter, derived for each latitude from the component of the earth's rotation in the direction of the local vertical. This leads to an arbitrary yet qualitative summary of the various scales of motion as shown in Table 5.2 using the Rossby number as an argument. Concerns with regional air quality suggest that the space and time scales of importance are those defined by a Rossby number greater than 1.

Table 5.2 gives an indication of the spacing required in general terms but does not indicate the optimum locations of monitoring stations for an individual situation. If cost were no object, the temporal and spatial resolution of the air quality could be measured to a very fine degree. Good spatial resolution is achieved by assuming that the signal-to-noise ratio is high and, with sensors at a large number of locations, good temporal resolution is achieved by sampling over short intervals for a long duration (Bibbero and Young, 1974). Cost, however, is a real factor and one is faced with using a limited number of sensors to achieve the most meaningful data for the problem at hand. All networks have the general objective of maximum valid data capture for the most reasonable cost.

A major dilemma that faces the monitoring of the transport and diffusion of air pollutants is that most meteorological phenomena, such as clear air turbulence, become more serious as the forcing functions become stronger. This leads to a definable if not tractable problem at the extreme. In marked contrast, public concern about an air-pollution situation increases rapidly as the transport and/or diffusion processes decrease in intensity, reaching an extreme where these values become near zero. This extreme situation is least tractable from the scientific viewpoint, in that the regional scale forcing functions become ill-defined or non-existent and small relatively disorganised local effects take over. In other words, an air pollution monitoring network must be designed to account for extremely localised influences and, at the same time, maintain a reasonable representativeness, so that interpolations can be made between adjacent stations.

An obvious consideration in the design of a sampling network is the need to arrange stations such that observations will be able to reproduce as closely as possible the overall pattern of pollution concentration, yet not miss highly localised effects.

As Munn (1970) has noted, network design criteria depend upon the purpose of the investigation, the nature of the underlying surface and such practical limitations as accessibility, availability of electric power and protection from vandalism. All of these factors must be accounted for, but the purpose of the investigation is the primary consideration in determining the scale of motion. Too close a spacing may introduce noise, and it is then often desirable to use a smoothing technique. Too wide a spacing, on the other hand, may result in partial loss of signal and aliasing.

The purpose of the investigation affects not only the initial design but is also critical in defining the optimum network. Measurements of peak concentrations from isolated sources will of necessity have different requirements than measurements of the overall background concentration.

Noll and Miller (1977) stress the importance of knowing the areas where maximum concentrations are likely to occur and state that these areas should be used as sensor locations. This obviously requires an understanding of the prevailing meteorological conditions and would lead to the establishment of a variety of stations for particular meteorological conditions. Hence, for each major pollutant source a number of downwind stations corresponding to the predominant wind directions would need to be established. If all maximum concentrations are to be covered then stations would be required for most compass directions.

In practice, a compromise must be reached between the amount of data collected (that is, the number of sites) and the cost. Noll and Miller (1977) recommend that the number of stations can be reduced to an optimum by estimating the probability of recording maximum concentrations (within a tolerance of ± 10 per cent) at a given station and omitting those with a low probability. Although this method indicates maximum concentrations, the resulting network may not be appropriate for interpolating regional air quality between stations.

In estimating the station density, Munn (1973) has recommended the calculation of correlation coefficients for each element under consideration and the construction of isopleths of equal correlation around some reference station. This is essentially a simplification of the method originally outlined by Drozdov and Sepelevskij (1946) and applied by Gandin (1965, 1970) and Sneyers (1973) for estimating synoptic station densities through the statistical structure of meteorological fields. The structure refers to the laws which the field obeys on average; that is, the regularities observed in large aggregates of data rather than in the behaviour of individual values. Hence, conclusions regarding the statistical structure depend on how the average is obtained and on the volume of observations, which could theoretically be infinite.

To have stable statistics it is necessary to have intervals between successive moments of time such that the individual data, collected at these moments, are not strongly correlated (Gandin, 1970). Due to the overriding annual cycle and the daily cycle in pollutant levels, such requirements are not easily met.

The approach of Gandin (1970) employs some restrictive assumptions that make it inappropriate for air-quality monitoring networks. Nevertheless, his method uses the standard error of interpolation between two stations and the structure function of the element under consideration. It yields an optimum network spacing for a given error in interpolation between stations. A similar criteria was employed by Toussaint (1980), who used a Kalman filter technique to select the optimal location of his sensors on the basis of minimum error. This approach has the advantage of defining the optimal locations for a given number of sensors without evoking the restrictive assumptions of Gandin (1970) but requires a detailed knowledge of the local meteorology.

The use of correlation coefficients on the other hand (Munn, 1973) identifies stations that are providing duplicate information or, alternatively, those where large spatial gradients may be incompletely observed. Thus, the existing network can be modified to rectify these conditions.

Both this technique and that of Gandin (1970) require a knowledge of the underlying field, which can only be obtained from measurements. Hence we are led to the concept of a dynamic network, as envisaged by Munn (1970), in which the initial network is modified as information regarding the structure of the pollution field becomes available.

Other attempts to estimate the optimum network spacing involve mathematical modelling of the transport and diffusion of pollutant (Langstaff *et al.*, 1987). As such, they suffer from the inherent inaccuracies of these mathematical models. Seinfeld (1972) developed a criteria that the concentration data collected at the stations be as sensitive as possible to changes in the emissions from major polluting sources, whereas Nakamori *et al.* (1979) opted to estimate a measurement system which gave representative concentrations for a certain urban area. Despite the different criteria for an optimal network both of these approaches require detailed simulation of the local meteorology.

Sheih *et al.* (1978) further suggest that the network design criteria should be specified in terms of the verification requirements of numerical models for atmospheric transport and dispersion. This network must be able to detect the general concentration pattern, the maximum concentrations, the background levels and have sufficient stations downwind to detect possible chemical transformations.

The initial design of an air pollution network is generally by trial and error, combined with some physical insight and all studies continue with a 'guided error' approach. Although there is a natural tendency to attempt to organise stations at equidistant grid points, Landsberg (see Munn, 1970) recommends that stations should be in lines at right-angles to the main physical, biological and cultural features. The spatial variation in the number of receptors should also be accounted for and this may lead to a clumping of stations in a particular region.

Once the network is established, the dynamic quality of the sensor locations should be recognised. As information becomes available, station locations should be modified such that the station separation is determined by the concentration gradient. The use of correlation coefficients between stations is a particularly powerful tool in this regard.

Alternatively, if the meteorology of a particular region is known, methods such as those outlined by Toussaint (1980) have the advantage that they can give an estimate of the optimum locations for a given number of sensors. Most other methods need to specify the concentration field and so in a sense the optimal criteria are determined by the knowledge of the field from the existing measurements. Obviously, if these measurements are inadequate a correlation coefficient analysis will point to areas of weakness but will not be able to specify the optimal locations.

The number of stations required under a particular optimal criteria may not be constant from season to season or even month to month. This is clearly illustrated in Table 5.3 with regard to the number of stations required for measuring monthly mean concentrations of sulphur dioxide in Nashville (Stalker *et al.*, 1962). In the summer months the number of

Table 5.3 Number of sampling stations required for measuring two-hourly, daily, monthly and annual mean concentrations of certain atmospheric pollutants in Nashville, Tennessee, with ± 20 per cent accuracy at the 95 per cent confidence level

Time period		Dustfall	Suspended particulate matter	Sulphur dioxide
2-Hourly		–	–	38
Daily		–	9	38
Monthly	January	100	2	10
	February	16	2	10
	March	22	3	11
	April	18	3	5
	May	20	2	1
	June	11	2	1
	July	20	2	1
	August	57	7	1
	September	27	5	1
	October	25	3	10
	November	80	4	9
	December	41	2	8
Annual		7	2	–

Source: Stalker *et al.*, 1962.

stations required is minimal as concentration levels are very low (near the sensor threshold), whereas in the winter an increased number of stations is required to specify the spatial variability across Nashville.

This table also indicates that there is a disparity in the number of stations required for different sampling intervals and emphasises the need to specify clearly the purpose of the network before adopting any criteria to assess it.

All of the above methods only consider the spatial location of the network and obviously similar considerations must be addressed in determining the temporal resolution of a network. In particular, the rate of sampling must be chosen to ensure adequate data retrieval. If too short a sampling interval is chosen the data may be swamped by noise. Alternatively, if too long a sampling interval is chosen the data may be aliased and meaningful variations such as daily or weekly oscillations may be lost (Johnson and Ruff, 1975; Crutcher, 1984).

Also, it has been assumed that all stations are sited so as to be representative of their local environment. Any station that is under the influence of purely local phenomena and not representative of the overall objectives of the network should be clearly identified from its lack of correlation with neighbouring stations. Such stations should be resited, unless local influences are important, in which case it may be necessary to establish additional stations.

The ultimate value of any network revolves about the data returned from the network. A high rate of valid data capture is imperative if the network is to fulfil its goal and in this a comprehensive preventative maintenance programme is essential. The optimal design of the network is pointless unless all stations achieve as close to 100 per cent data capture as possible.

The maintenance programme should also incorporate routine calibration of all sensors following the lines outlined by Johnson and Ruff (1975). In this way the most efficient use can be made of the limited resource and an optimal air-quality network can become a reality.

5.2 Meteorological monitoring

The meteorological instrumentation required to characterise the boundary layer has been reviewed by Lenschow (1986) and Stull (1988) amongst others, whereas the instrumental requirements for air-quality studies have been discussed by Mason and Moses (1984), Johnson and Ruff (1975) and Hoffnagle *et al.* (1981) and is summarised in Tables 5.4 to 5.7. These illustrate that the central requirements of an air-quality assessment are the systematic evaluation of the following four items:

(i) The horizontal wind field measured at a number of stations depending on the complexity of the terrain and the horizontal variability of the wind field. These measured wind-speed and direction data may be interpolated across a uniform grid to simulate the horizontal variation of the wind field. Cup or propeller anemometers must be accurate to within 0.2 ms^{-1} ± 5 per cent of the wind speed, with a start speed of less than 0.5 ms^{-1} and a distance constant (63 per cent recovery) of less than 5 m. Wind vanes must have a resolution of 1° and an accuracy of 5°. Delay distance (50 per cent recovery) must be less than 5 m with a damping ratio of greater than 0.4. Sixty or more samples will estimate hourly means to within 5–10 per cent. Sample averaging time should be 1–5 s, with a response time of 1 s. At least three-hundred-and-sixty samples are required to estimate the hourly deviation within 5–10 per cent (Hoffnagle *et al.*, 1981).

(ii) The atmospheric stability which partly controls the rate of dispersion and mixing of effluents discharged into the atmosphere. Atmospheric stability is computed from the vertical temperature gradient, or the fluctuations in wind direction, or through consideration of global solar radiation, wind speed and dry bulb temperature. The measurement of temperature gradients requires special attention with proper shielding and aspiration, and at least 0.1°C accuracy and 0.02°C resolution. The

Table 5.4 Estimated importance of relevant variables according to scale of interest and elevation of emissions source

	Microscale (0–1 km)		Local scale[§] (0–10 km)		Mesoscale[§] (10–100 km)		Regional scale (100–1,000 km)	
Variable	Surface	Elevated	Surface	Elevated	Surface	Elevated	Surface	Elevated
Near-surface wind direction	1	3	1	2	4	4	4	4
Vertical wind-direction profile	3	1	2	1	1	1	1	1
Near-surface wind speed	1	3	1	2	4	4	4	4
Vertical wind-speed profile	3	1	2	1	1	1	1	1
Horizontal wind-speed profile	4	4	3	3	2	2	1	1
Atmospheric stability	4	2	2	2	3	3	4	4
Vertical diffusivity	2	2	2	2	4	4	4	4
Horizontal diffusivity	2	2	2	2	2	2	2	2
Mixing depth	5	4	3	3	2	2	2	2
Near-source aerodynamic effects	2	2	3	3	5	5	5	5
Surface conditions	3	4	4	4	5	5	5	5
Topographic effects	3	2	3	2	3	3	4	4
Solar radiation	4	2	2	2	3	3	4	4
Solar UV radiation*	5	5	4	4	2	2	3	3
Precipitation[†]	1	1	1	1	1	1	1	1
Temperature and humidity[‡]	5	5	4	4	3	3	3	3

Note: 1 = most important; 5 = least important.

* Assuming pollutant is photochemically generated.

† Assuming pollutant is particulate in nature.

‡ Assuming pollutant is subject to chemical transformations.

§ The urban scale (5–50 km) overlaps the local mesoscales.

Source: after Johnson and Ruff, 1975

Table 5.5 Typical applications of direct sensors

Meteorological element	Symbol	Typical direct sensors
Near-surface mean wind speed	$\overline{V}_0$	Mechanical anemometer (cup, propeller)
Near-surface wind-speed fluctuations	V'_0	Sensitive mechanical anemometer (cup, propeller); hot-wire anemometer; sonic anemometer
Near-surface mean wind direction	$\bar{\theta}_0$	Wind vane; multicomponent propeller anemometer
Turbulent fluctuations of near-surface horizontal and vertical wind directions	θ'_0; α'_0	Sensitive wind vane; bivane; sensitive multi-component propeller propeller anemometer
Wind-velocity profile	$V(z)$	Tower-mounted anemometers and vanes; pilot balloons tracked by theodolite; balloons positioned by radar, navigational aids or Doppler radio techniques
Horizontal air trajectory	$x,y(t)$	Constant-level balloons (tetroons), positioned by transponder-aided radar or other radio techniques
Vertical temperature gradient	$\Delta T/\Delta z$	Differential thermometers (thermocouple, resistance-wire)
Temperature profile	$T(z)$	Tower-mounted thermometers; balloon-borne radiosondes (free-rising or tethered)
Dewpoint/relative humidity	DP	Wet-bulb thermometers; lithium chloride strips and other humidity-sensitive elements; dewpoint hygrometers
Solar radiation	SR	Pyranometers
Solar UV radiation	UV	Solar photometers
Net radiation	NR	Net radiometers
Turbidity/visual range	β	Transmissometer; nephelometer; visual observation
Precipitation	P	Rain/snow gauges (tipping bucket, weighing, etc.)
Cloud cover	C	Celiometer; all-sky camera; visual observation

Source: after Johnson and Ruff, 1975.

Table 5.6 Determination of important variables in air-pollution meteorology

Process	Relevant variable	Elements required for measurement	Elements required for estimation
Transport	Wind speed and direction (near-surface and profile)	$\overline{V}_0$, $\bar{\theta}_0$, $V(z)$	–
↓	Horizontal air trajectory (Lagrangian)	$x(t)$, $y(t)$ of moving balloon	$V(x, y, t)$ at z of interest
Diffusion	Eddy diffusivity	–	V', θ', α'
	Gaussian diffusion coefficients	Tracer concentration distribution	Atmospheric stability aerodynamic effects
	Atmospheric stability	–	Various combinations of $\Delta T/\Delta z$, $\Delta V/\Delta z$, $\sigma_o{}'$, $\overline{V}_o$, SR, NR, C
↓	Mixing depth	$T(z)$, DP(z), $V'(z)$ $\beta(z)$, $T'(z)$	T_0 at time of interest plus t $T(z)$ at some earlier time
Transformation	Air physical state	$T(z)$, DP(z), $\beta(z)$	–
	Precursor travel time	–	Horizontal air trajectory (see above)
↓	Solar UV radiation	UV(z, t)	$(SR)_0$, β_0, mixing depth
Removal	Precipitation scaverging	Precipitation sampler	$P(x, y, t)$; horizontal air trajectory
↓	Dry deposition	Dustfall plate	$V'(x, y, z)$; horizontal air trajectory

Source: Johnson and Ruff, 1975.

Table 5.7 Summary of some stability classification techniques

Classification technique	Key variables	Weakness
Richardson number	$(\Delta T/\Delta z)/(\Delta V/\Delta z)^2$	Function of height; required measurement accuracy for vertical gradients is hard to achieve
Bulk stability parameter	$(\Delta T/\Delta z)/V^2$	Same as above
Delta – *T* classification	$\Delta T/\Delta z$	Same as above
Sigma – θ classification	Δ_θ	Wind meander and surface conditions not taken into account
Pasquill/Turner (Turner, 1964)	*V*, *C*, time of day	Surface conditions and thermal advection not taken into account

Source: Johnson and Ruff, 1975.

time constant (63 per cent recovery) of the temperature probe should be one minute with comparison probes having identical response characteristics. The sample averaging time should be greater than or equal to half the time constant (Hoffnagle *et al.*, 1981). The average must be a true average before any digital recording is attempted.

(iii) Mixed layer height which may be maintained convectively by surface heating or mechanically by wind-generated turbulence. This height may be obtained continuously from a monostatic acoustic sounder which determines the height of any rapid changes of temperature gradient or wind shear from the two-way travel time. It is usual to correlate the continuous acoustic sounder data with irregular temperature profiles, either by tethered sonde (i.e. Lyons *et al.*, 1982) or aircraft-borne sensors. Acoustic-sounder resolution is about 10 m, and the useful range varies from 50 to 1,000 m. Since the mechanically generated mixed layer normally lies below 200 m, it is best measured from an instrumented tower supported by the acoustic sounder (Hoffnagle *et al.*, 1981).

(iv) The turbulent diffusion coefficients. These may be derived from the atmospheric stability and distance from the source using Pasquill–Gifford stability classification. Hoffnagle *et al.* (1981), however, emphasise the preference for on-site measurements of turbulence intensity to characterise the dispersion properties.

The direct sensors required for these measurements are summarised in Table 5.5 and the features of some remote sensors are summarised in Table 5.8. In general it is imperative that the sensors operate for at

Table 5.8 Features of various remote sensors as applied to air pollution meteorology

Parameter	Sensor	Main operating principle*	Category					
			Potential for all weather operation	Tracer availability	Spatial resolution	Time resolution	Accuracy	Annoyance and/or safety
Temperature and/or stability	Acoustic	Backscatter from thermal fluctuations	B	A	B	B	C	B£
	Laser	Raman backscatter from nitrogen	B	B	A	B	B	B
	Microwave radiometer	Emissions from 50–60 GHz O_2 band	A	N/A	B to C	B	B	A
	Radio-acoustic sounding system (RASS)	Tracking of acoustic wave by CW Doppler radar	B	A	B	B	B	B
	Acoustic	Bistatic Doppler; angle of arrival of acoustic return	B	A	B	B	B	B†
	Laser	Spatial correlation of aerosol fluctuations	B	B	A	A	A	B
	Radar	Pulsed dual Doppler	C	C	A	A	A	A
Mixing layer depth	Acoustic	Backscatter from thermal fluctuations	B	A	A	A	A	B†
	Laser	Backscatter from aerosols	B	A‡	A	A	A	B
	Radar FM/CW	Backscatter from refractive index changes	A	B	A	A	A	A

* Added to Beran and Hall's original table
† Changed from an A as appeared in the original table
‡ Changed from a B as appeared in the original table.

Source: Adapted from Beran and Hall, 1973. A = most favourable; C = least favourable.

least one year to establish a representative climatology of the region, although recent studies suggest that data should be collected over at least five years to ensure a representative year is chosen for analysis. Most current Environmental Impact Studies tend to collect data over one year to establish climatological trends or to identify the contribution of the proposed project to the background concentrations. In addition, worst case studies are conducted to identify the maximum ground-level concentration that is likely to be experienced. For example, in the case of an industrial complex located near the coast, the highest ground-level concentration is likely to be experienced under fumigation conditions associated with the sea breeze. Hence it is important to be able to specify the meteorology of the event to account for the fumigation concentration, as well as have sufficient data to assess the overall contribution to the climatological background concentration.

5.3 Pollutant monitoring

Knowledge of the concentration of atmospheric pollutants is obviously required to validate and calibrate atmospheric dispersion models and for monitoring long-term trends in air quality. In the calibration process it is normal practice to use a tracer, that is, a pollutant with no natural or artificial sources in the region, whose emission can be controlled. Normal industrial emissions come and go and are generally difficult to characterise. In this manner, quantitative agreement can be sought between the modelled and measured concentrations and the model tuned. Actual industrial plumes have been used to quantify dispersion parameters (Gifford, 1980; Yassky, 1983; Johnson, 1983) but this tends to be in the validation of diffusion formulae rather than the calibration of a model. Various tracer techniques have been reviewed by Johnson and Ruff (1975) and Johnson (1983), whereas the techniques for the measurement of atmospheric pollution have been summarised by Anonymous (1978) and Cadle (1975), and reviewed by Katz (1977, 1980) and Ferrari and Johnson (1984).

In the measurement of pollutants it is important to account for:

(i) the time and space scale of interest;
(ii) the type of pollutant (such as O_3, NO_x, PAN, hydrocarbons, aerosols);
(iii) the expected level (see Table 5.9);
(iv) the calibration of the sensor.

If one is interested in a plume a few hundred metres wide, one does not sample from a single ground-based site: either an array of samplers is used or a mobile (van or aircraft) sampling procedure is followed. Likewise,

Table 5.9 Rough concentrations of pollutants expected in the atmosphere

	Background adds about 10%	Urban (concentrated)
Aerosols	1–10 $\mu g\ m^{-3}$	100 $\mu g\ m^{-3}$
Numbers	1,000	100,000 cm^{-3}
SO_2	100 $\mu g\ m^{-3}$ (smog formation)	100 $\mu g\ m^{-3}$ (stratosphere)
O_3	50 $\mu g\ m^{-3}$	50 $\mu g\ m^{-3}$
NH_3	5 $\mu g\ m^{-3}$	< 5 $\mu g\ m^{-3}$
Nitrogen oxides	.3 ppm (N_2O)	100 ppm (burning at high temperature)
CO	.1 ppm	10 ppm
HC	1 ppm (CH_4)	10 ppm

sampling the plume from an aircraft flying at, say, 100 ms^{-1}, requires an instrument that responds in less than a second to acquire a reasonable resolution of the plume. Alternatively, if one is interested in climatic variations in aerosol concentration, one does not gather data at one-second time intervals, filling rooms with spools of magnetic tape.

Also, ordinary constituents may be pollutants. Water can be considered a pollutant when it acts as a catalyst to oxidize iron or when it undergoes a reaction such as $NO_2 + H_2O \rightarrow H_2NO_3$, which produces an undesirable product. In the home we have many materials that are produced by our activities (cooking, smoking, painting) that are health risks. In addition, the natural background levels of Radon gas may be very high; some 10,000 homes in the United States are said to have unacceptable levels of Radon decay products (Nero, 1988). These contaminants are every bit as harmful as industrial pollutants and motor-car emissions. Natural materials are not necessarily healthy materials.

Back to the point: in measurement, computers cannot be relied on to do all the work, including the statistics (see Keith, 1987). Before, during and after making the measurements, it is most important to arrange routine calibration checks (Taylor, 1987). Without this the data obtained are virtually worthless; all instruments drift or breakdown; regular and irregular blank samples should be submitted for analysis; calibration checks with standards are essential.

Wet chemistry techniques are useful when the samples are collected on filters, through sampling trains or in plastic bags. Using this method one simply proceeds with a standard spot-test technique or a conventional analysis by, say, measuring properties of distilled water that has been used to elude the pollutants from the filter. Standards are produced by comparisons with the results from solutions of known concentration (Harrison and Perry, 1986).

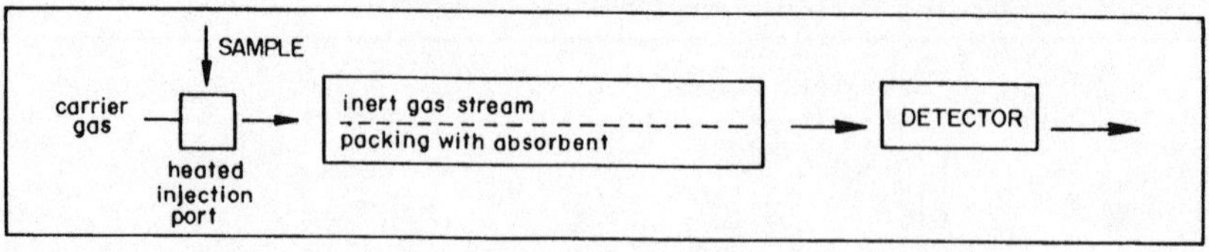

Figure 5.2 Gas chromatography

A variation of this procedure is the electrochemical procedure used to measure SO_2 and O_3. In a polluted situation, SO_2 may be detected by bubbling through a solution of H_2O_2. The H_2O_2 oxidizes the SO_2 to $SO_4^{=}$ and the amount collected is detected by measuring the increase in electrical conductivity of the solution. O_3 is detected by measuring the amount of current required to produce just enough hydrogen to remove iodine formed from KI by O_3. Another variation is the standard spot-testing vials used by industry to detect small quantities of pollutants. The gas is simply drawn through a disposable tube packed with solid absorbant that changes colour when it reacts with, say, CO_2.

The Drager tester, for example, utilises a small hand-operated bellows pump to draw the sample through the tube and, depending on the level of pollutant, one simply draws more or less sample air through the tube. Of course, the final concentration is obtained by dividing by the volume of air sampled. The technique is, however, insensitive at levels below 1 ppm.

Sometimes, of course, more sophistication is needed. Currently, this might involve gas or HPLC chromatography, atomic absorption or emission spectroscopy.

Gas chromatography consists of a column packed with an inert material coated with an absorbent. The technique uses differential partitioning between the liquid and gas phases as shown in Figure 5.2. The sample is injected into an inert gas carrier (N_2, He or H_2) that passes through the column. The absorbent material is selected to have special absorbent properties for specific chemical groups. Molecules of slightly different absorption properties are partitioned differently on the absorbing surface and absorb and desorb to differing degrees. The result is a separation of the sample into its components with each component arriving at the column exit at differing times. Low-boiling compounds generally pass through quickly and higher-boiling compounds may take very long times to emerge (perhaps half an hour or more). The effluent may be detected in a number of ways including:

(i) thermal conductivity;
(ii) flame ionization;

(iii) electron capture;
(iv) flame spectroscopy.

The thermal conductivity detector consists of two temperature sensitive elements positioned, respectively, in the sample gas stream and a reference stream. As the pollutant gas passes the detector, the different thermal conductivity creates a temperature difference that is a measure of the presence of the pollutant. Since the thermal conductivity is proportional to the collision cross-section and inversely proportional to the square root of molecular weight, the system is of general utility, provided the unknown has sufficient volatility. Sensitivity is about 1 ppm so the technique is most useful in polluted situations.

The flame ionization detector burns the gaseous effluent from the column in a hydrogen-oxygen flame. The impurity substances (hydrocarbons or C-H groups) create ions in the flame, which are detected as an increased flame conductivity. The technique can detect 0.003 ppm propane and hence is valuable at ordinary atmospheric levels of pollutants.

The electron-capture detector is sensitive to the presence of certain substances by their ability to capture thermal electrons. A low energy β emitter, such as ^{63}Ni or ^{3}H, is used as a source and capture results in a decrease in the electrical conductivity of the gas because of the low mobility of the ions that are formed. The detector responds to molecules containing halogens or oxygen atoms and molecules with conjugated double bands. It can detect 10^{-5} ppm SF_6 gas or 5×10^{-3} ppm PAN and hence is a good low-level atmospheric monitor.

High Performance Liquid Chromatography (HPLC) is a modern variation of the chromatographic technique (Skoog, 1985, Chapter 7; Lim, 1986). It has all the basic components of chromatographic assemblies including a sampling system/elution system and a detection system. Rather than a gas carrier, however, a liquid at high pressure is used. Pressures may be as high as 6,000 psi to provide relative high eluent flow rates. Fine packaging materials with particles of diameter 3–10 μm are used to realize rapid separation of the components of a sample. The technique is suitable for separating non-volatile species or fragile ones, including amino acids, proteins, nucleotides, hydrocarbons, carbohydrates, drugs, pesticides, hormones, metal-organic species and a variety of inorganic substances.

Ion chromatography is similar to the HPLC technique but features ion exchange resins as separating columns (Fritz, 1982). Resins are either Cation-exchanging or Anion-exchanging. In cation exchange, hydrogens are usually replaced by cations and these are then exchanged for hydrogen ions from the eluent, following separation along the column. In anion exchange, chloride ion is often exchanged for anions during separation along the column. At least one column of either type is required for simultaneous determination of the cations and anions in a sample. Detection is usually by measuring the electrical conductivity of the effluent.

Atomic absorption uses the specific atomic absorption properties of atoms, wherein light of a selected wavelength is passed through the sample and the amount of absorption is measured.

Usually the sample is dispersed in a flame or otherwise separated into its elements with intense heat. Aerosols, for instance, are trapped in a carbon boat. The boat is heated with electrical current and the emitted vapours analysed by atomic absorption. The light used is ordinarily chopped and sophisticated electronic techniques are used to maximise the signal-to-noise ratio. The technique is capable of detecting elements, particularly metals, below the ppb range. The main disadvantage of the technique is the necessity for a relatively large sample. For example, if cadmium is present at, say, the 0.01 per cent level in the aerosol sample and the air sample contained 100 μm m^{-3}, 1 m^{-3} of air would only yield 0.0001 gms of solid. This, perhaps, could be analysed if added to 1 ml of H_2O, making the solution 0.01 ppm Cd with the detection limit as 0.005 ng. The problem is that it is difficult to collect a sample of 1 m^{-3} in less than, perhaps, half an hour. Secondly, atomic absorption is incapable of analysis of chemical components, such as $SO_4^{=}$ and any number of other groups.

Emission spectrometry is similar to atomic absorption except that a light source is not used. Some atoms, particularly sodium, the alkalis and alkaline earths, strongly emit light of a particular wavelength when heated. One simply adds the sample to a flame and observes the emission – with sodium, this is an intense orange colour. The technique can detect sodium well below the ppb range, and in particular, can detect calcium at the 0.1 ppb level. A variation on this technique called scintillation flame photometry, detects the flashes of individual particles as they pass through the flame. You may have noticed the light 'twinkles' that occur in a portable gas flame at a beach party. These are emissions by single salt-containing particles, the sodium emitting its distinctive orange colour. Electronic counting techniques have been developed that allow the detection and counting of particles as small as 0.03 μm. This corresponds to a mass of 7×10^{-17}.

Fluorometry is a form of emission spectrometry that uses light as the exciting source. The emissions are then detected at, perhaps, 90° to the exciting light source. Emissions are usually at longer wavelengths than the exciting wavelength, in the range 300 to 800 nm. Many organic molecules exhibit strong fluorescence. In fact, the technique is not sufficiently specific and usually needs to be combined with a pre-separation procedure, perhaps using a chromatograph. Atoms also exhibit fluorescence and, in fact, zinc can be detected at the 0.04 ppb level with atomic fluorescence techniques.

The ring-oven technique is, perhaps, the simplest and least expensive of the techniques that may be mentioned. This is a variation of the spot-test techniques that have evolved from testing scrapings from paintings by the

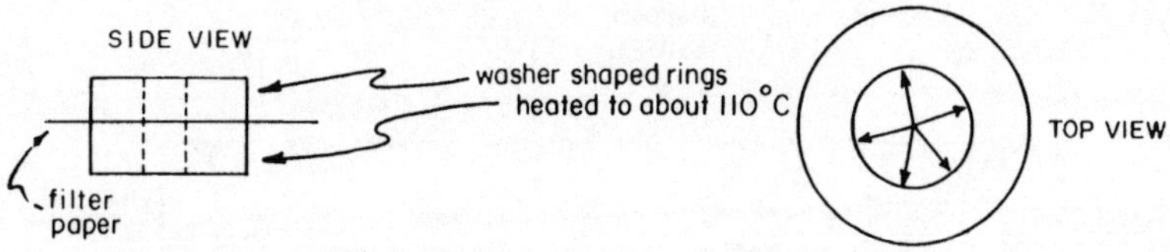

Figure 5.3 Ring-oven technique

old masters. It simply involves collecting the sample on a piece of appropriate filter paper and placing the sample in a ring-shaped oven consisting, effectively, of two washers placed on either side of the paper as shown in Figure 5.3. The washers are heated to above the boiling point of the appropriate solvent, usually water. Then the solvent (distilled water) is dripped in μ drops in the centre of the paper. The water is absorbed outward (see arrows) and evaporates near the hot metal rings. Of course all solubles in the sample are also drawn to the rings. On evaporation of the solvent, the trace substances are left in a tight ring near the metal rings. The tight rings are simply analysed by wet, trace detection procedures. The ring oven concentrates the material more than 100 fold. Generally, the analysis can detect sub-microgram quantities and it is often sensitive to nanogram quantities. It is good for nearly any element or chemical group. For example, the test for lead is sensitive to 0.01 μg. The tests normally produce a red ring with this element and standardization can give a quantitative accuracy of about ± 30 per cent. The technique is particularly good for $SO_4^{=}$, NH_4^{+}, and complex organic groupings for which tests are available. Its main attribute is a small sample requirement, perhaps only a nanogram (see Weisz, 1970).

Infra-red absorption spectrometry can be used to detect specific chemical bonds but, usually, requires concentrations above the ppm level. Nuclear Magnetic Resource utilises atomic particles to irradiate the sample and produce unstable isotopes. The technique can be sensitive at concentrations less than 1 ppm. Mass spectrometry is generally sensitive to concentration of the order of 1 ppm. It has the advantage of looking at individual species, including activated forms and, has been used to detect concentrations of rare, radical species at the 0.1 ppb level.

Lasers have considerable potential for remote detection of trace gases and aerosols. Two procedures are possible – forward scattering or back scattering – with possible frequency alteration due to Raman, chemical effects. Sensitivities of the order of 1 ppb should be possible in 1 km paths.

Correlation spectrometry is a proven approach that measures the concentration integrated over the path between the light source and the detector. One practical approach uses the sun or skylight as the source of light. The

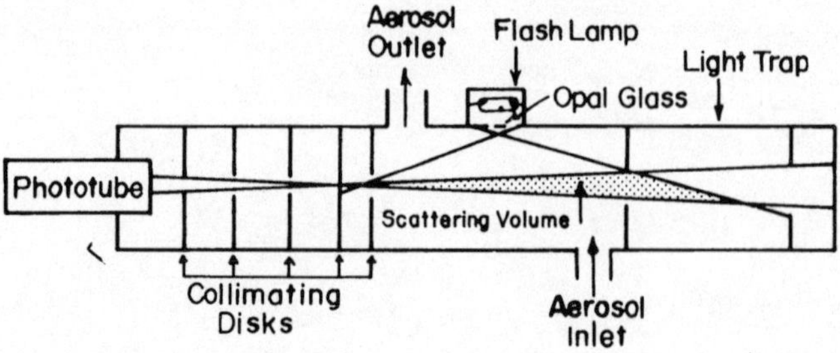

Figure 5.4 The integrating nephelometer (after Butcher and Charlson, 1972)

trace material in the beam absorbs particular portions of the solar radiation. Inside the instrument the absorption peaks are compared by vibrating the incoming beam at an audio frequency. Then the light is separated into its wavelength components by an optical grating and the output spectrum compared with a prepared spectral 'correlation mask'. The technique is particularly useful for detection of SO_2 and NO_2 and can detect 10 ppm in a 100 m path. It also can be used for large-scale studies and is particularly effective mounted in a mobile van.

Aerosol sampling can be sophisticated or simple. One can simply collect the largest particles on a Frisbee (Hall and Upton, 1988) but in a sophisticated research project it might happen that scanning electron microscope techniques would be used to look at the structure of submicron particles or an electron microprobe might be used to obtain their atomic constitution. That does not mean the simple techniques are not adequate; in fact, often the statistics of the measurement can be improved with multiple samplings even though the analysis is relatively qualitative.

One measuring system that is now of general use is the integrating nephalometer. It consists of a long tube at the centre of which is located an intense flash tube that by means of opaque glass emits a cosine distribution of light intensity with angle. The light is scattered by the sample through nearly 0 to 180° scattering angles (see Figure 5.4). A sensitive photomultiplier tube senses the scattered light specifically through a series of colliminating holes, giving an integrated 0–180° effect, which tends to smooth out difficulties with complex scattering. Synchronization of the flash with the electronic sensing circuitry gives an extreme signal-to-noise ratio that even allows detection of molecular scattering.

The electrical mobility analyser utilises a measurement of electrical mobility (velocity of a particle of unit charge in a unit electric field) as an indication of size (see Figure 5.5). Particles are charged negatively in a prescribed manner. They are then placed in a cylindrical, annular space surrounding an envelope of clear air which, in turn, has a positively

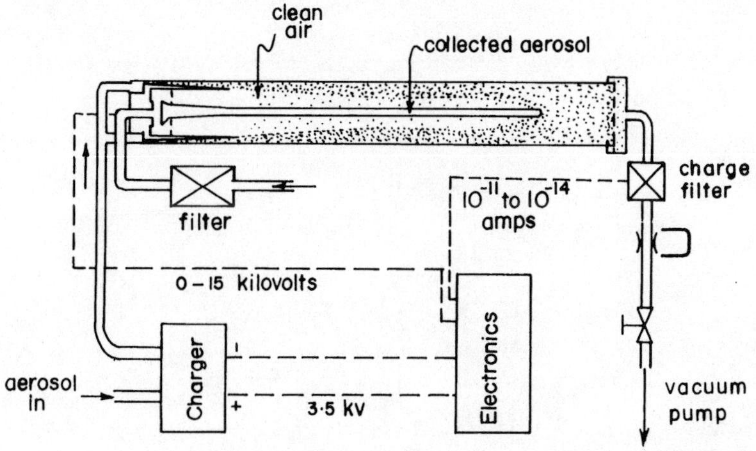

Figure 5.5 The electric mobility analyser (after Husar, 1974). Copyright ASTM. Reprinted with permission.

charged electrode at its axis. Depending upon the flow (residence time) and the applied voltage, the negatively charged particles are collected on the positively charged axial rod. The lowest voltages collect the particles of the highest mobilities; the highest voltages, those with the lowest mobilities. The final sensing is done by collecting the charged particles that remain in the original sample with a current filter, a filter connected to a sensitive current electrometer.

There are basically three types of Condensation Nuclei Counters, the Aitken particle counter, the Pollak counter and the continuous GE counter. The first of these, the Aitken counter was invented by John Aitken in the early 1900s when he discovered the myriad of small particles less than 0.1 μm which today bear his name. The counter utilises the cloud-chamber concept, which creates a moist, highly supersaturated (~ 300 per cent) environment of water vapour by a large expansion of the air. This large excess of moisture condenses on every particle whose size is above that of small ions, nucleating droplet formation. The droplets rapidly grow to visible size and a photograph of the cloud is taken with an intense flash of light. The optical depth and viewing area are set so that the effective sample volume is known. Results are obtained by simply, though tediously, counting the droplet images on the film. The instrument is a primary standard.

A later version, developed by Pollak (Pollak and Metnieks, 1960) through a lifetime of painstaking attention to detail, utilises the principle of light scattering and extinction. This instrument consists of a long tube with a wetted inner wall. Air is drawn through the tube, the sample sealed off with a series of valves, clean air is added to the sample to create a sufficient

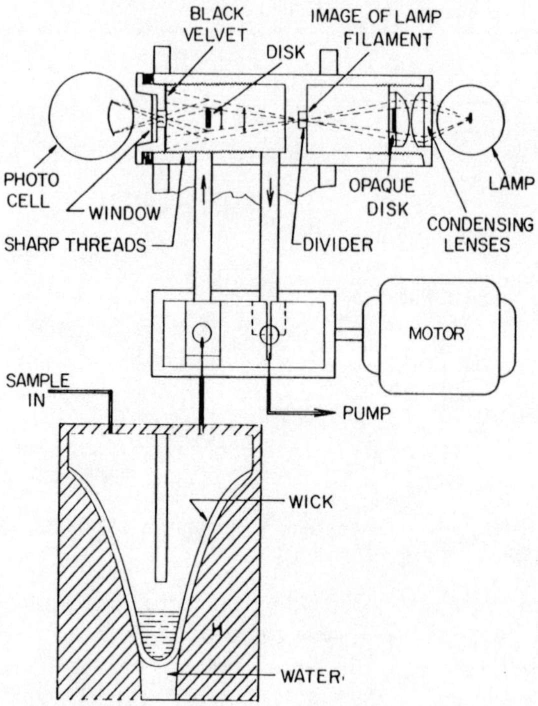

Figure 5.6 Diagram of GE condensation nuclei counter

positive pressure, the sample allowed time to become saturated and then expanded by the release of a quick-action toggle valve. This sampling procedure is similar to that used in an Aitken counter.

However, the particles are not individually counted with the Pollak counter; they are detected by measuring the extinction of a collimated light source. The source is placed at the top of the counter; an electronic photosensor (selenium cell) detects the intensity of light reaching the other, lower end of the tube. The calibration in terms of particle numbers is given by Pollak, provided one strictly follows his design. The instrument is considered a secondary standard.

The continuous Aitken counter or GE counter utilises the same principle with a different geometry and a smaller sample size (see Figure 5.6). In addition, the instrument is automated so that it continuously takes samples, expands them and measures the light extinction. Whereas the other counters generally require several minutes to take a sample, the GE counter samples every 0.5 seconds. It is not as stable as the Pollak or Aitken counter nor is it as sensitive (its lowest detection limit is around 100 particles cm^{-3}

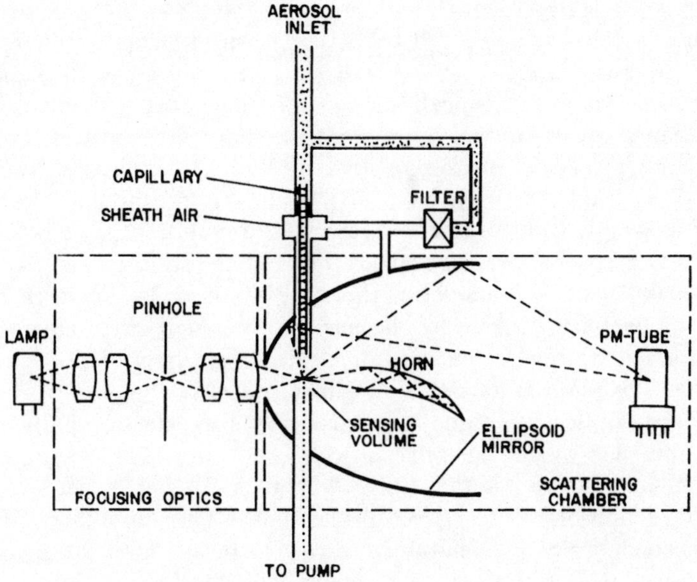

Figure 5.7 The ellipsoid mirror optical particle counter (from Husar, 1974). Copyright ASTM. Reprinted with permission.

whereas the Pollak counter can detect levels of around 20 particles cm^{-3}). However, it gives a resolution sufficient for aircraft or other mobile sampling systems. In many cases it is the backbone instrument used for pollution estimates and source location.

The Single Particle Optical Counter draws the sample through a small glass capillary tube a few millimetres in diameter. Light is focused on the axis of the capillary and, as particles pass through the tube, the light beam is affected as each individual particle passes the beam. Light collection optics and a sensitive, fast photomultiplier tube supply a voltage signal to counting electronics. The amplitude of the signal is related to particle size and fast electronic pulse height analysers can give a direct read-out of particle numbers in different size ranges, i.e. a particle size distribution. There are many different optical configurations, the 90° or side-angle viewing sensors being the least sensitive. The complexity of the scattering, however, makes side-angle viewing ambiguously affected by particle shape. The best arrangement appears to be the ellipsoidal mirror which gathers light from all angles and is little affected by particle shape (see Figure 5.7). It can reliably measure particles in the size range of 0.3 to 10 μm.

The electron microscope is a valuable detector not only for counting and sizing but also for characterising the particles in terms of shape and chemical properties. Two types of microscope are common, the transmission electron microscope and the scanning electron microscope (SEM). The transmission microscope is the least expensive and generally one views the 'shadows' of the particles, which are suspended on a thin plastic film on a minute grid of copper. Measurements down to 0.01 μm are common with elaborate sputtering processes and chemical reaction procedures being used to detect the third dimension or the chemical composition of the particles.

The scanning electron microscope uses a beam to actually scan the surface and get a topographical map of the surface and the particles lying upon it. The sample generally needs to be coated with a metallic film and, when properly prepared, present-day electron systems are capable of resolutions below 0.01 μm. A variant on the system utilises the electron beam (with varying energies) to activate the particle to emit X-rays. This system (the electron microprobe) is capable of elemental analysis of the particles through these emissions down to the element oxygen.

The ordinary optical microscope is invaluable for detecting particles of specific shape such as cubical salt particles or fibrous, asbestos particles. The high volume sampler is the ordinary sampler for particulate mass. It uses a large suction blower to collect all the aerosol mass on a large fibrous filter; subsequent weighting, elution and wet chemistry give chemical and physical analyses. The modern counterpart of this device is the quartz crystal piezoelectric microbalance which detects the mass of aerosol directly by measuring the frequency at which the crystal oscillates. Other standard particulate-measuring devices include the paper tape sampler (which collects the sample directly on a thin roll of paper tape), the directional dust gauge (which consists of four cylinders with open faces directed towards the four points of the compass), as well as the cascade impactor, which collects the aerosol in known particulate fractions (Katz, 1977).

5.6 Pollutant effects on plants

Atmospheric pollution can also be monitored by its impact, such as the increased rate of corrosion that is evident in some urban areas or through its effect on vegetation. Although there is clear evidence of pollutant damage to vegetation (i.e. Jacobson and Hill, 1970; McLaughlin, 1985; Mathy, 1987) and a decrease in species diversity or indicator plants such as lichens has been noted in polluted areas, many authors have attempted to relate pollutant damage to actual airborne concentrations.

O'Gara (1922) derived the law of gas action on a plant cell and showed that for short-term fumigations,

$$(\chi - \chi_r)\, t = k,$$

where χ is the airborne concentration, t is the exposure time, χ_r is the

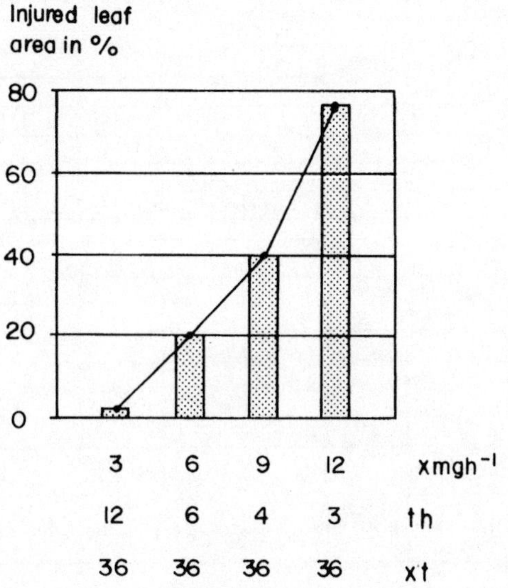

Figure 5.8 Progressive increase in the degree of injury of radish with an increase in SO_2 concentration (after van Haut, 1961)

threshold concentration at which no injury occurs even under long-term exposures and k is a constant depending on the level of injury. Obviously, as the airborne concentration decreases towards the threshold concentration, the exposure time for injury to occur will approach infinity.

The law of gas action was applied by Thomas and Hill (1935) in investigating the effects of SO_2 and, under conditions of extreme plant sensitivity, they found:

$(\chi - 0.24)t = 0.94$	slight necrosis of leaves
$(\chi - 1.40)t = 2.10$	50 per cent necrosis
$(\chi - 2.60)t = 3.20$	100 per cent necrosis

where χ is in mg h^{-1} and t is in hours. Such a relationship implies that foliar damage is directly proportional to the product of concentration and exposure time. Hence, low concentrations over a long period would have the same effect as high concentrations over a much shorter period. More recent studies (van Haut, 1961) have shown that damage induced by SO_2 does not follow this threshold law, but that foliar damage increases with the concentration, even when the product, χt, is held constant, as shown in Figure 5.8. Guderian *et al.* (1960) suggest that the effects of high SO_2 concentration can be accounted for by

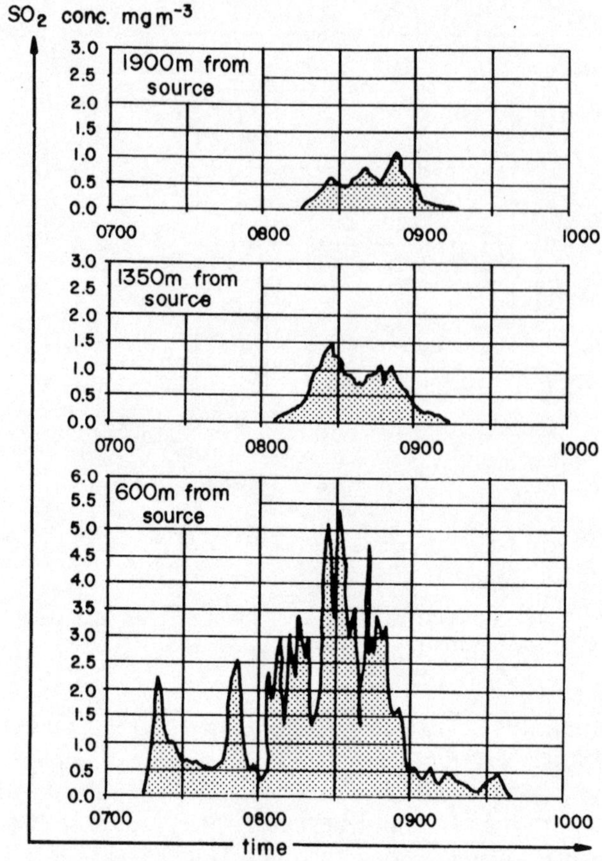

Figure 5.9 Example of SO_2 concentration from a single source as a function of distance from the source (after Guderian, 1977)

$$t - t_r = k \exp(-a(\chi - \chi_r)), \qquad (5.2)$$

where t_r is the threshold time, that is the minimum time necessary to cause injury, a is a parameter representing the complex internal and external growth factors and k is the growing time. Under this equation products of χt are dependent on χ and will decrease as the concentration increases, in contrast to O'Gara's law of gas action.

The effects of SO_2 which show an increase with the concentration cannot be explained by an increase in sulphur accumulation in the plant, as it has been shown that sulphur accumulation from the uptake of SO_2, within a certain range of concentrations, increases more with exposure time

Table 5.10 Effect of exposure to nitrogen dioxide on pulmonary function

	NO_2 exposure concentration ($\mu g\ m^{-3}$)	(ppm)	Study population	Effect
Los Angeles: high-exposure group				
Median hourly NO_2	130	0.069	205	No difference in
90th percentile NO_2	250	0.133	office	most tests.
Median hourly O_x		0.15	workers	Smokers in both
90th percentile O_x		0.15		cities showed
				greater changes
San Francisco: lower-exposure group				in pulmonary
Median hourly NO_2	65	0.035	439	function than
90th percentile NO_2	110	0.058	office	non-smokers.
Median hourly O_x		0.02	workers	
90th percentile O_x		0.03		

Source: after Linn *et al.*, 1976.

than with concentration (Guderian, 1977). Effects that increase with the increased ambient SO_2 concentration result from a higher uptake rate rather than a higher absolute pollutant uptake. Hence, per unit of time, less time is available for detoxification of the pollutant in the plant under high ambient concentrations (Ziegler, 1975). If little of the pollutant is absorbed per unit of time, oxidation of the sulphite to sulphate reduces the toxicity thus lessening the incidence of injury (Thomas *et al.*, 1944 a,b).

The concentration and duration of action of a pollutant at a single receptor will change constantly through the action of atmospheric turbulence. Near a single source the concentration distribution (Figure 5.1) is particularly peaked and changes from very high concentrations to pollutant-free periods will be observed under slight changes in wind direction. Obviously, as you move further from the source, the exposure time to a pollutant as a function of the total time will be further reduced (Figure 5.9). Thus the effect of intermittent pollutant action needs to be considered if the effects on plants are to provide a quantitative measure of pollutant concentrations.

Equations 5.1 and 5.2 were developed for the continuous action of pollutants and do not allow for possible plant recovery during times of no pollutant. Guderian (1970 a,b,c) investigated this effect by exposing winter barley to SO_2 at a concentration of 4.1 mg m^{-3} of air. The pollutant load in terms of concentration and total duration of exposure were held constant but the pollutant-free periods varied for each experiment. Exposure was once for 2 hours (1 × 2), twice for 1 hour (2 × 1), four times for a half

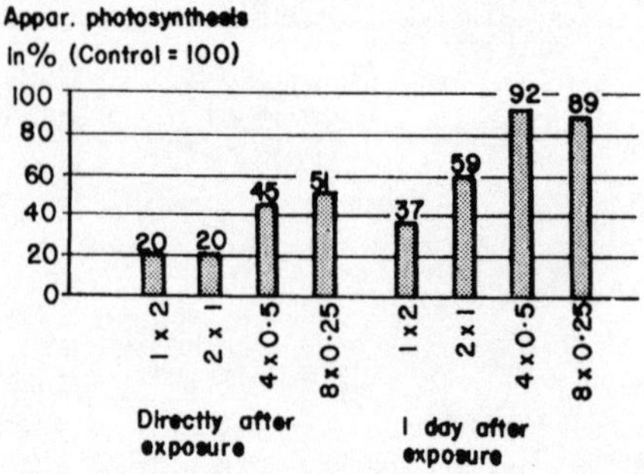

Figure 5.10 Apparent photosynthesis of winter barley dependent on frequency and length of exposure to SO_2 at a constant concentration (after Guderian, 1977)

hour (4 × 0.5) and eight times for a quarter hour (8 × 0.25) with equivalent pollutant free periods between exposures. During the exposures (1 × 2) and (2 × 1), the CO_2 uptake fell to 20 per cent of the control, whereas photosynthesis during the (4 × 0.5) and (8 × 0.25) exposures was only reduced by about 50 per cent. On the following day, without additional exposure, plants under the (4 × 0.5) and (8 × 0.25) regimes had experienced almost total recovery to normal photosynthesis, whereas the other plants were exhibiting more gradual recovery as shown in Figure 5.10 (Guderian, 1977).

These experiments indicate that recovery can occur during pollutant-free periods provided that at a particular concentration a definite exposure duration has not been exceeded, as suggested by the law of gas action. Thus, although plant damage or a decrease in species diversity can indicate the presence of pollutant, they do not give a clear quantitative estimate of pollutant concentration.

5.5 Pollutant effects on humans

The health effects of air pollutants, especially for people with respiratory problems, have been known for some time. However, except in exceptional circumstances, such as the Los Angeles smog (Schoettlin and Landau, 1961), there is often no clear cause and effect relationship between a particular airborne pollutant and a corresponding health disorder. This is hardly surprising, given the complexity of the human situation. Nevertheless, a variety of studies (such as Burgess *et al.*, 1973; Dockery and Spengler, 1981; Mainwaring, 1989) have been made of the effects of particular pollutants and these have been reviewed by Ferris (1978).

Table 5.11 Effects of sulphur dioxide and particulate matter on human health

SO_2 ($\mu g\ m^{-3}$)	SO_2 (ppm)	Suspended particulates ($\mu g\ m^{-3}$)	Effects
Short-term effects			
250	0.095	350	Increased respiratory symptoms in patients with chronic bronchitis
722	0.276	350	No change in pulmonary function of patient with chronic lung disease
200–300	0.076–0.114	230	Decreased forced expiratory volume
200	0.076	150	Increased frequency of asthma attacks
Long-term effects			
250	0.095	250	Increased phlegm production
130	0.05	240	Increased respiratory disease
120	0.046	180	Increased respiratory illness and decreased pulmonary function
120	0.046	230	Increased lower respiratory illness
55	0.021	180	Increased respiratory symptoms, decreased pulmonary function
37	0.014	131	No effect
66	0.025	80	No effect

Source: after Ferris, 1978; Wadden and Scheff, 1983.

Table 5.12 Effects of ozone, O_3, or oxidant, O_x, exposure

Concentration (ppm)	Exposure duration (hours)	Averaging time	Pollutant	Effects	Reference
0.01–0.3		Hourly average	O_3	Lung function parameters in 25 per cent of school children tested were significantly correlated with O_3 concentration in the two hours prior to testing	Kagawa and Toyama (1975)
0.1–0.15		Daily maximum hourly average	O_x	Increased rates of respiratory symptoms and headache were reported by students on days when O_x concentrations exceeded 0.15 ppm compared to days when O_x concentrations were less than 0.10 ppm	Makin and Mizoguchi (1975)
0.25	2,4		O_2	No lung function changes observed in 'reactive' subjects (those who had histories of respiratory problems associated with air pollution) while performing intermittent, light exercise	Hackney *et al*. (1975a, b, c)
0.37	2		O_3	Exposure to O_3 and SO_2 together produced changes in lung function substantially greater than the sum of the separate effects of the individual pollutants	Hazucha and Bates (1975)
0.37	2		O_3	Smaller effect than observed by Hazucha and Bates (1975) suggests previous study simulated smog episode in regions having high oxidant and sulphur pollution	Bell *et al*. (1977)

Of particular concern are the health effects associated with the common urban pollutants of NO_2, SO_2 and O_3 and these are summarised in Tables 5.10, 5.11, and 5.12. Most health effects associated with nitrogen oxides are attributed to NO_2 and levels of this gas above 282 mg m^{-3} (150 ppm) can be lethal while levels between 94 and 282 mg m^{-3} (50–150 ppm) can produce chronic lung disease (Ferris, 1978). Studies on the health effects of SO_2 generally consider particulates also to incorporate the combination of SO_2 particulates and sulphate and these are summarised in Table 5.11.

Ozone is a pulmonary irritant that affects the mucous membranes, other lung tissues and respiratory function. High urban concentrations are normally produced as a result of the effect of sunlight on nitrogen oxides and hydrocarbons emitted from automobile exhausts and other combustion processes. Health effects of ozone and oxidants are summarised in Table 5.12.

5.6 Pollution indices

So far we have concentrated on the estimation of pollutant concentrations and gained some insight into the impact of these concentrations on both plant and human life. Yet it is useful to have a single pollution index that could account for both the expected level of pollutant and its overall impact. Munn (1970) has classified the various proposals for such an index in the diagram shown in Figure 5.11. The philosophy behind the choice of an individual index is discussed by Munn (1975) and various indices have been reviewed by Ott and Thom (1976) and Ott (1978). Lag times between levels 1 and 2 in Figure 5.11 are usually of the order of hours, and air-quality stress is generally predictable on the basis of our understanding of atmospheric diffusion. The lag time between air-quality stress and ecological response is variable and not well understood. Hence an air-pollution index based on level 1 or 2 data will reflect day-to-day variations, whereas an index based on level 3 data will tend to reflect more long-term trends in air quality (Munn, 1970).

Green (1966) developed an index based on two pollutant variables, the concentration of sulphur dioxide and the coefficient of haze. The coefficient of haze is a reading obtained by passing air through a paper tape and measuring the reduced light transfer that results. As such, it is a measure of the particulate concentration in the atmosphere. Green's index was defined as

$$I = \tfrac{1}{2}(I_1 + I_2) \qquad (5.3)$$

with

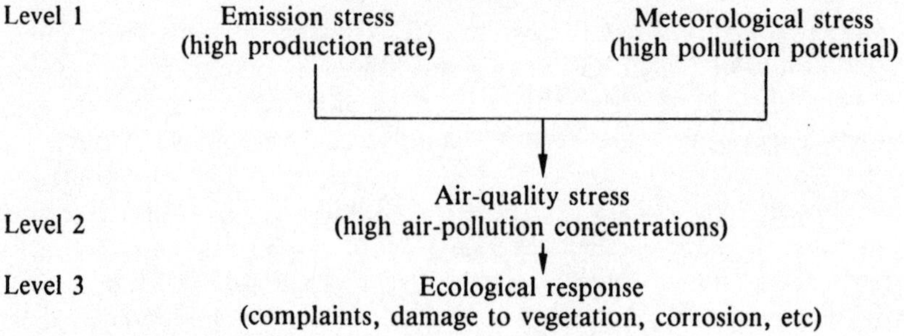

Figure 5.11 Air pollution indices (after Munn, 1970)

$$I_1 = 84\ \chi^{0.431}$$
$$I_2 = 26.6\ Y^{0.576}$$

where χ is the concentration of SO_2 in ppm and Y is the coefficient of haze.

Since the index is based on level 1 data as defined in Figure 5.11, it was designed as a system for triggering control actions during air-pollution episodes. As defined, the index has three main levels: 'desired level' (I = 25), 'alert level' (I = 50) and 'extreme level' (I = 100), at which various control strategies can be initiated. This index is limited to sulphur dioxide and particulates and suffers from what Ott (1978) defines as eclipsing. That is, extremely poor environmental quality may exist for one pollutant variable but this will not be reflected in the overall index as it is a weighted sum of two variables.

An alternative index, the combustion products index, was developed by Rich (1967) as an overall air-pollution index for a region and is defined as

$$I = \frac{F}{V} \tag{5.4}$$

where F is the quantity of fuel burned in the area and V is the ventilation volume. Assuming a mixing depth of h over an urban area of crosswind width w, the ventilation volume for a mean wind speed u is given by

$$V = u\ h\ w$$

and this represents the volume of air into which the pollutants are emitted. Such an index is analogous to the box model (equation 3.47) but cannot account for pollutant types or changing meteorological conditions and, in particular, atmospheric stability.

In 1970, Toronto introduced an air-pollution index to give warning of, and to prevent, the adverse effects of a build-up of air pollution which may occur during prolonged periods of stagnant weather (Shenfeld, 1970). Like that of Green, this index concentrated on the concentration of sulphur dioxide and the coefficient of haze and was defined as

$$I = 0.2\,(I_1 + I_2)^{1.35} \tag{5.5}$$

with

$$I_1 = 126\,\chi$$
$$I_2 = 30.5\,Y$$

where the pollutant variables, χ, Y are 24-hour running average values (i.e. 24 1-hour values were averaged together, one beginning at each hour of the day). A value of this index less than 32 was considered as an acceptable air quality, whereas the values in excess of 50 constituted an alert situation and 100 is regarded as an air pollution episode threshold.

Further refinements of air-pollution indices to incorporate more pollutants (Babcock, 1970) and various air-quality objectives (Inhaber, 1974) have been developed and a pollutant standards index (Thom and Ott, 1976) proposed. This index is a segmented linear function based on the United States National Ambient Air Quality Standards and it reports only the highest numerical value of all sub-index values for each pollutant. It has been described in detail by Ott (1978).

All of these air-quality indices give a measure of air-quality stress but do not give a full measure of the environmental impact of that air-quality stress; that is, the individual index is not defined at level 3 of Figure 5.11.

5.7 Environmental assessment

The failure to define a single index that can account for the full environmental impact of air-borne pollutants is hardly surprising given the complexity of atmospheric diffusion as well as the need to incorporate value judgements in the final assessment of damage (see Figures 5.12, 1.1). Nevertheless, the Gaussian models and their variants used in regulatory applications provide the framework for an overall assessment of expected ambient concentrations. It is important to remember that this is only the beginning of an environmental assessment, and the consequences of the pollutant need to be considered in terms of the damage or otherwise that may result. In this instance, damage is defined to include all effects that reduce the intended value or use of the environment, be that economic, ecologic or aesthetic. Obviously such considerations involve the incorporation of social value judgements and deliberate decisions about acceptable tradeoffs (cf.

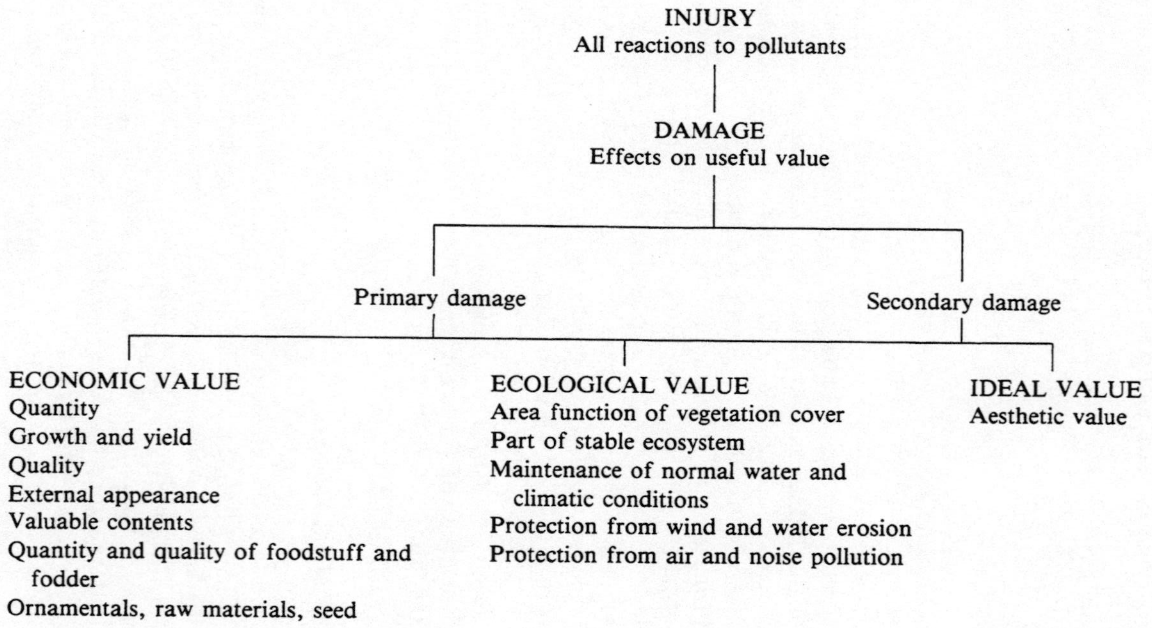

Figure 5.12 Environmental impact of pollutants to vegetation (after Guderian, 1977)

O'Riordan and Wynne, 1987; Renn, 1989), which are beyond the scope of this text, but are nevertheless an essential component of the assessment.

Environmental impact assessment should address how the impact will be caused, what magnitude and range of impact will occur, the significance of the impact in comparison to known standards and the methodology used to identify the impact (Zeller, 1984). The first requirement is thus a knowledge of the proposed emission strength as well as the existing emissions within the region. The strength of emission may suggest a possible environmental hazard by itself or alternatively, combined with other emissions in the region, it could result in an air-quality stress. If there is likely to be significant diurnal variations in the source these should also be accounted for.

Having defined the source terms, the second requirement is to specify the dispersion potential and the local meteorology. Depending on the location, air-quality stress could occur over short or long periods. For example, a coastal site could lead to high ground-level concentrations under fumigation conditions as well as unacceptable long-term concentrations. In this case, the environmental impact study should address the likely occurrence of the short-term episode conditions when fumigation will result in high ground-level concentrations in the vicinity of the source as well as the climatological or annual background concentrations in the broader region. The monitoring required for each of these is determined by the scale of the phenomena (Table 5.4), so different monitoring and modelling strategies may need to be adopted.

The definition of expected concentrations requires a justification of confidence in the modelling strategy. Thus an environmental impact assessment should be able to defend the methodology adopted by comparison with tracer studies that provide an inherent calibration on the models used, or alternatively, through reference to published results elsewhere. This gives an indication of the relative errors involved in the concentration estimates and enables an estimate of the degree of confidence in the calculated concentrations and their trends.

The accurate assessment of air quality trends is an essential first step towards defining the overall environmental assessment. Munn (1975) has addressed the broader issues in outlining the requirements for the assessment of environmental impact through a consideration of the ecosystem as a whole, whereas Jakeman and Simpson (1983) have applied a risk assessment strategy to air-quality management in the Hunter Valley region of New South Wales. This essentially takes the philosophy of using the most practicable means of assessing the likely air quality from existing and proposed sources. By identifying areas where a combination of either meteorological or emission stress may lead to unacceptable air quality, alternative policy strategies can be developed to mitigate or lessen the impact of these emissions. Such strategies can vary from reduced emissions under certain atmospheric conditions, to changes in land use zoning or changes in transport policy. Either way a validated air-quality model can provide an assessment of the net outcome of alternative strategies and define the most feasible in terms of acceptable risk. Of course, such models need to be updated as more information in terms of our understanding of the environment or basic observational data becomes available, so as to establish the foundations of a dynamic environmental management system. The success of such a system is heavily dependent on the accurate assessment of the pollutant pathways, which is the fundamental goal of air-pollution meteorology.

APPENDIX: STANDARD AIR-QUALITY MODELS

The following briefly describes standard air-quality models that are available in computer compatible form from the US Environmental Protection Authority. Many of these models are finding routine application in environmental impact studies and forming the basis of regulatory air-quality modelling (Sawford and Ross, 1985).

RAM (Turner and Novak, 1978 a,b)

This short-term Gaussian steady-state algorithm estimates concentrations of stable pollutants from urban point and area sources. Hourly meteorological data are used, and hourly concentrations and averages over a number of hours can be estimated. The model uses Briggs plume rise and Pasquill–Gifford dispersion equations with dispersion parameters for urban areas. Concentrations from area sources are treated by assuming that sources directly upwind are representative of area source emissions affecting the receptor. Special features include the determination of receptor locations downwind of significant sources and the determination of locations of uniformly spaced receptors to ensure good area coverage with a minimum number of receptors.

CRSTER (Anonymous, 1977)

Algorithm estimates ground-level concentrations resulting from up to nineteen collocated elevated stack emissions for an entire year and prints out the highest and second-highest 1-hour, 3-hour and 24-hour concentrations as well as the annual mean concentrations at a set of 180 receptors (five distances by thirty-six azimuths). The algorithm is based on a modified form of the steady-state Gaussian plume equation which uses empirical dispersion coefficients and includes adjustments for plume rise and limited mixing. Terrain adjustments are made as long as the surrounding terrain is physically lower than the lowest stack-height input. Pollutant concentrations for each averaging time are computed for discrete, non-overlapping time periods (no running averages are computed) using measured hourly values of wind speed and direction, and estimated hourly values of atmospheric stability and mixing height.

CDM (Busse and Zimmerman, 1973)

The climatological dispersion model determines the long-term (seasonal or annual) quasi-stable pollutant concentrations at any ground level receptor using average emission rates from point and area sources and a joint frequency distribution of wind direction, wind speed and stability for the same period.

HIWAY (Zimmerman and Thompson, 1975)

This model computes the hourly concentrations of non-reactive pollutants downwind of roadways, and is applicable for uniform wind conditions and level terrain.

VALLEY (Burt, 1977)

Algorithm is a steady-state, univariate Gaussian plume-dispersion model designed for estimating either 24-hour or annual concentrations resulting from emissions from up to fifty (total) point and area sources. Calculations of ground level pollutant concentrations are made for each frequency designated in an array defined by six stabilities, sixteen wind directions, and six wind speeds for 112 program-designated receptor sites on a radial grid of variable scale. Empirical dispersion coefficients are used and include adjustments for plume rise and limited mixing. Plume height is adjusted according to terrain elevations and stability classes.

PAL (Petersen, 1978)

This short-term Gaussian steady-state algorithm estimates concentrations of stable pollutants from point, area and line sources. Computations from area sources include effects of the edge of the source. Line source computations can include effects from a variable emission rate along the source. The algorithm is not intended for application to entire urban areas but for smaller-scale analysis of such sources as shopping centres, airports and single plants. Hourly concentrations are estimated and average concentrations from 1 hour to 24 hours can be obtained.

PTPLU (Pierce *et al.*, 1982)

PTPLU is a point-source Gaussian dispersion-screening model for estimating maximum surface concentrations for 1-hour concentrations. PTPLU is based on Briggs plume-rise methods and Pasquill–Gifford dispersion coefficients. It is an adaptation and improvement of PTMAX allowing for wind-profile exponents and other optional calculations, such as buoyancy-induced dispersion, stack downwash and gradual plume rise. PTPLU produces an analysis of concentration as a function of wind speed and stability class for both wind speeds constant with height and wind speeds increasing with height. Use of the extrapolated wind speeds and other options allows the model user a more accurate selection of distances to maximum concentration. PTPLUI is the interactive version of this model.

MPTER (Pierce and Turner, 1980)

MPTER is a multiple point-source Gaussian model with optional terrain adjustments. MPTER estimates concentration on an hour-by-hour basis for relatively inert pollutants (i.e. SO_2 and TSP). MPTER uses Pasquill–Gifford dispersion parameters and Briggs plume-rise methods to calculate the spreading and the rise of plumes. The model is most applicable for source–receptor distances less than 10 kilometres and for locations with level or gently rolling terrain. Terrain adjustments are restricted to receptors whose elevation is no higher than the lowest stack top. In addition to terrain adjustments, options are also available for wind-profile exponents, buoyancy-induced dispersion, gradual plume rise, stack downwash and plume half-life.

BLP (Schulman and Scire, 1980 a,b)

BLP (buoyant line and point-source dispersion model) is a Gaussian plume-dispersion model designed to handle the unique modelling problems associated with aluminium reduction plants, and other industrial sources where plume rise and downwash effects from stationary line sources are important.

ISCST (Bowers *et al.*, 1979 a,b)

The industrial source complex short-term model is a steady-state Gaussian plume model which can be used to assess pollutant concentrations from a wide variety of sources associated with an industrial source complex. This model can account for settling and drying deposition of particulates, downwash, area, line and volume sources, plume rise as a function of downwind distance, separation of point sources, and limited terrain adjustment. Average concentration or total deposition may be calculated in 1-, 2-, 3-, 4-, 6-, 8-, 12- and/or 24-hour time periods. An '*N*'-day average concentration (or total deposition) or an average concentration (or total deposition) over the total number of hours may also be computed.

ISCLT (Bowers *et al.*, 1979 a,b)

The industrial source complex long-term model is a steady-state Gaussian plume model which can be used to assess pollutant concentrations from a wide variety of sources associated with an industrial source complex. This model can account for settling and dry deposition of particulates, downwash, area, line and volume sources, plume rise as a function of downwind distance, separation of point sources, and limited terrain adjustment.

ISCLT is designed to calculate the average seasonal and/or annual ground-level concentration or total deposition from multiple continuous point, volume and/or area sources. Provision is made for special discrete *x, y* receptor points that may correspond to sampler sites, points of maxima or special points of interest. Sources can be positioned anywhere relative to the grid system.

MPTDS (Rao, 1982; Rao and Satterfield, 1982)

MPTDS is a modification of MPTER to account explicitly for gravitational settling and/or deposition loss of a pollutant. Surface deposition fluxes can be printed under an optional output feature. MPTDS is a multiple point source code with an optional terrain adjustment feature. The code is primarily based upon MPTER, which has Gaussian modelling assumptions. Execution is limited to a maximum of 250 point sources and 180 receptors.

Hourly meteorological data are required and the period of simulation can vary from one hour to one year.

PALDS (Rao, 1982; Rao and Snodgrass, 1982)

PALDS calculates hourly concentrations from point, area and line sources. PALDS is a modification of PAL and has the capability to explicitly treat the effects of gravitational settling and/or deposition loss of pollutant on calculated concentrations. This is an optional feature. Surface deposition fluxes are also printed as output under this option. PALDS reads all input as data cards. Source input includes point, area, horizontal line, special line, horizontal curved path and special curved path data. All source input data are optional. Meteorology and receptor input data are also required.

SHORTZ (Bjorklund and Bowers, 1982 a,b)

SHORTZ is designed to calculate the short-term pollutant concentration produced at a large number of receptors by emissions from multiple stack, building and area sources. SHORTZ uses sequential short-term (usually hour) meteorological inputs to calculate concentrations for averaging times ranging from one hour to one year. The model is applicable in areas of both flat and complex terrain, including areas where terrain elevations exceed stack-top elevations. The program requires random-access mass storage capability.

LONGZ (Bjorklund and Bowers, 1982 a,b)

LONGZ is designed to calculate the long-term pollutant concentration produced at a large number of receptors by emissions from multiple stack, building and area sources. LONGZ uses statistical wind summaries to calculate long-term (seasonal or annual) average concentrations. The model is applicable in areas of both flat and complex terrain, including areas where terrain elevations exceed stack-top elevations. The program requires random-access mass storage capability.

MESOPUFF (Bass *et al.*, 1979; Benkley and Bass, 1979b)

MESOPUFF is a variable-trajectory regional-scale Gaussian puff model especially designed to simulate the air-quality impacts of multiple point sources at long distances. Highly user-oriented, MESOPUFF provides a range of flexible options. It is designed to be driven by user-specified meteorological scenarios, of arbitrary duration, constructed by a suitable meteorological pre-processor, MESOPAC. It outputs spatially-gridded concentration arrays averaged over arbitrary time intervals of one hour or more and is designed to be coupled to a post-processor, MESOFILE, to provide additional graphical and statistical analyses. Routines are provided for: plume rise, plume growth, fumigation, linear conversion of SO_2 to SO_4, and dry deposition of SO_2 to SO_4.

MESOPLUME (Bass *et al.*, 1979; Benkley and Bass, 1979a)

MESOPLUME is a mesoscale plume segment (or 'bent plume') model designed to calculate concentrations of SO_2 and SO_4 over large distances. Highly user-oriented, MESOPLUME provides a range of flexible options. It is designed to be driven by user-specified meteorological scenarios, of arbitrary duration, constructed by a suitable meteorological pre-processor, MESOPAC. It outputs spatially-gridded concentration arrays averaged over arbitrary time intervals of one hour or more and is designed to be coupled to a post-processor, MESOFILE, to provide additional graphical and statistical analyses. Routines are provided for: plume rise, plume growth, fumigation, linear conversion of SO_2 to SO_4, and dry deposition of SO_2 to SO_4.

ROADWAY (Eskridge and Thompson, 1982)

ROADWAY is a finite difference model which predicts pollutant concentration near a roadway. It uses surface layer similarity theory to produce vertical wind and turbulence profiles. Temperatures at two heights and wind velocity are required inputs. These values are usually obtained from instruments on a tower upwind of the roadway. ROADWAY and ROADCHEM use the vehicle-wake theory (Eskridge and Hunt, 1979) as modified and verified in wind-tunnel experiments (Eskridge and Thompson, 1982) to predict velocity and turbulence along the roadway.

ROADCHEM (Eskridge and Thompson, 1982)

ROADCHEM is a version of ROADWAY incorporating chemical reactions involving NO, NO_2 and O_3 as well as advection and dispersion of NO, NO_2, O_3 and CO. It uses surface-layer similarity theory to produce vertical wind and turbulence profiles. Temperatures at two heights and wind velocity are required inputs. These values are usually obtained from instruments on a tower upwind of the roadway.

PTMAX (Turner and Busse, 1973)

Performs an analysis of the maximum short-term concentrations from a single-point source as a function of stability and wind speed. The final plume height is used for each computation. Uses Briggs plume rise and Pasquill–Gifford deposition methods to estimate hourly concentrations for stable pollutants. PTMAXI is the interactive version of this model.

PTDIS (Turner and Busse, 1973)

Estimates short-term concentrations directly downwind of a point source at distances specified by the user. The effect of limiting vertical dispersion by a mixing height can be included and gradual plume rise to the point of final rise is also considered. An option allows the calculation of isopleth half-widths for specific concentrations at each downwind distance. Uses Briggs plume rise and Pasquill–Gifford deposition methods to estimate hourly concentrations for stable pollutants. PTDISI is the interactive version of this model.

PTMPT (Turner and Busse, 1973)

Estimates the concentration for a number of arbitrarily located receptor points at or above ground-level, from a number of point sources. Plume rise is determined for each source. Downwind and crosswind distances are determined for each source-receptor pair. Concentrations at a receptor from various sources are assumed additive. Hourly meteorological data are used: both hourly concentrations and averages over any averaging time from one hour to twenty-four hours can be obtained. Uses Briggs plume rise and Pasquill–Gifford deposition methods to estimate hourly concentrations for stable pollutants. PTMTPI is the interactive version of this model.

REFERENCES

Al-Nakshabandi, G. and H. Kohnke (1965). Thermal conductivity and diffusivity of soils as related to moisture tension and other physical properties. *Agr. Meteorol.*, 2: 271–9.

Angstrom, A. (1916). Uber die Gegenstrahlung der Atmosphare. *Meteorol Z.*, 33: 529–38.

Anonymous (1972). *Safety guide for water cooled nuclear power plants.* US Atomic Energy Agency, Division of Reactor Standards, Safety Guide No. 23, 23.1–23. 13.

——— (1977). *User's manual for single source (CRSTER) model.* US Environmental Protection Agency, Research Triangle Park, N.C. EPA-450/2-77-013 (NTIS Accession Number PB-271 360).

——— (1978). *International operations handbook for measurement of background atmospheric pollution.* WMO No. 491, World Meteorological Organisation, Geneva, 110 pp.

——— (1989). Air quality modelling workshop, Parts I and II. *Clean Air*, 23.

ApSimon, H.M., J.J.N. Wilson and K.L. Simms (1989). Analysis of the dispersal and deposition of radionuclides from Chernobyl across Europe. *Proc. Roy. Soc.*, Series A, 425: 365–405.

Arritt, R.W., R.A. Pielke and M. Segal (1988). Variations of sulfur dioxide deposition velocity resulting from terrain-forced mesoscale circulations. *Atmos. Environ.*, 22: 715–23.

Atwater, M.A. and R.J. Londergan (1985). Differences caused by stability class on dispersion in tracer experiments. *Atmos. Environ.*, 19: 1045–51.

Babcock, L.R. (1970). A combined pollution index for measurement of total air pollution. *J. Air Poll. Control Assn.*, 20: 653–9.

Bagnold, R.A. (1941). *The physics of blown sand and desert dunes.* Wiley, New York, 265 pp.

Baker, D.G. (1965). Factors affecting soil temperature. *Minn. Farm Home Sci.*, 22: 11–13.

Baldocchi, D.D., B.B. Hicks and P. Camara (1987). A canopy stomatal resistance model for gaseous deposition to vegetated surfaces. *Atmos. Environ.*, 21: 91–101.

Bass, A., C.W. Benkley, J.S. Scire and C.S. Morris (1979). *Development of mesoscale air quality simulation models: Volume 2. Comparative sensitivity studies of puff, plume and grid models for long-distance dispersion.* US Environmental Protection Agency, Research Triangle Park, N.C. EPA-600/7-80-058 (NTIS Accession Number PB80-227 580).

Batchelor, G.K. (1953). The conditions for dynamic similarity of motions of a frictionless perfect gas atmosphere. *Q. J. R. Meteorol. Soc.*, 79: 224–35.

Bell, K.A., W.S. Linn, M. Hazucha, J.D. Hackney and D.V. Bates (1977). Respiratory effects of exposure to ozone plus sulfur dioxide in southern Californians and eastern Canadians. *Am. Ind. Hyg. Assoc. J.*, 38: 696–706.

Bencala, K.E. and J.H. Seinfeld (1979). An air quality model performance assessment package. *Atmos. Environ.*, 13: 1181–5.

Benkley, C.W. and A. Bass (1979a). *Development of mesoscale air quality simulation models Volume 2. User's guide to MESOPLUME (mesoscale plume segment) model.* US Environmental Protection Agency, Research Triangle Park, N.C. EPA-600/7-80-057 (NTIS Accession Number PB80-227 598).

——— and ——— (1979b). *Development of mesoscale air quality simulation models Volume 3. User's guide to MESOPUFF (mesoscale puff) model.* US Environmental Protection Agency, Research Triangle Park, N.C. EPA-600/7-80-058 (NTIS Accession Number PB80-227 796).

Benarie, M.M. (1987). The limits of air pollution modelling. *Atmos. Environ.*, 21: 1–5.

Bibbero, R.J. and I.G. Young (1974). *Systems approach to air pollution control.* Wiley, New York, 531 pp.

Bierly, E.W. and E.W. Hewson (1962). Some restrictive meteorological conditions to be considered in the design of stacks. *J. Appl. Met.*, 1: 383–90.

Bjorklund, J.R. and J.F. Bowers (1982a). *User's instructions for the SHORTZ and LONGZ computer programs. Volume 1.* US Environmental Protection Agency, Philadelphia, Pa. EPA-903/9-82-004A (NTIS Accession Number PB83-146 092).

——— and ——— (1982b). *User's instructions for the SHORTZ and LONGZ computer programs. Volume 2.* US Environmental Protection Agency, Philadelphia, Pa. EPA-903/9-82-004B (NTIS Accession Number PB83-146 100).

Blackmore, D.R., M.N. Herman and J.L. Woodward (1982). Heavy gas dispersion models. *J. Hazardous Materials*, 6: 107–28.

Bluman, G. (1983). Dimensional analysis, modelling and symmetry. *Int. J. Math. Educ. Sci. Technol.*, 14: 259–72.

Boatman, M.L., C.C. Van Valin and D.L. Wellman (1988). Continuous atmospheric sulfur dioxide gas measurements aboard an aircraft: a comparison between the flame photometric and fluorescence methods. *Atmos. Environ.*, 22: 1949–55.

Bowers, J.F., J.R. Bjorklund and C.S. Cheney (1979a). *Industrial source complex (ISC) dispersion model user's guide. Volume 1.* US Environmental Protection Agency, Research Triangle Park, N.C. EPA-450/4-79-030 (NTIS Accession Number PB80-133 044).

———, ——— and ——— (1979b). *Industrial source complex (ISC) dispersion model user's guide. Volume 2.* US Environmental Protection Agency, Research Triangle Park, N.C. EPA-450/4-79-031 (NTIS Accession Number PB80-133 051).

Briggs, G.A. (1973). Diffusion estimation for small emissions. Env. Res. Lab., Air Resources Atmos. Turb. and Diffusion Lab., 1973 Annual Report, ATDL-106, USDOC-NOAA.

——— (1975). Plume rise predictions. In Haugen (1975), 59–111.

——— (1984). Plume rise and buoyancy effects. In Randerson (1984), 327–66.

——— (1985). Analytical parameterizations of diffusion: the convective boundary layer. *J. Climate Appl. Met.*, 24: 1167–86.

——— (1988). Analysis of diffusion field experiments. In Venkatram and Wyngaard (1988), 63–117.

Brunt, D. (1932). Notes on radiation in the atmosphere. *Q. J. R. Meteorol. Soc.*, 58: 398–420.

Bubenich, D.V. (1984). *Acid Rain Information Book*, 2nd edn. Noyes Publications, New Jersey, 397 pp.

Burgess, W., L. Di Berardinis and F.E. Speizer (1973). Exposure to automobile exhaust – III: An environmental assessment. *Arch. Environ Health.*, 26: 325–9.

Burt, E.W. (1977). *Valley model user's guide.* US Environmental Protection Agency, Research Triangle Park, N.C. EPA-450/2-77-018 (NTIS Accession Number PB-274 054).

Businger, J.A., J.C. Wyngaard, Y. Izumi and E.F. Bradley (1971). Flux-profile relationships in the atmospheric surface layer. *J. Atmos. Sci.*, 28: 181–9.

Busse, A.D. and J.R. Zimmerman (1973). *User's guide for the climatological dispersion model.* US Environmental Protection Agency, Research Triangle Park, N.C. EPA-R4-73-024 (NTIS Accession Number PB-227 346).

Butcher, S.S. and R.J. Charlson (1972). *An introduction to air chemistry.* Academic Press, New York.

Cadle, R.D. (1975). *The measurement of airborne particles.* Wiley Interscience, New York, 342 pp.

Carpenter, S.B., T.L. Montgomery, J.M. Leavitt, W.C. Colbaugh and F.W. Thomas (1971). Principal plume dispersion models: TVA power plants. *J. Air Poll. Control Assn.*, 21: 491–5.

Carras, J.N. and G.M. Johnson (1983). *The Urban Atmosphere – Sydney, A Case Study.* CSIRO Publications, East Melbourne, Victoria, 655 pp.

——— and D.J. Williams (1988). Measurements of relative σ_y up to 1800 km from a single source. *Atmos. Environ.*, 22: 1061–9.

Carson, J.E. (1961). *Soil temperature and weather conditions.* Rep. No. 6470, Argonne National Laboratories, Argonne.

——— and H. Moses (1962). The annual and diurnal heat-exchange cycles in upper layers of the soil. *J. Appl. Met.* (American Meteorological Society), 2: 397–406.

——— and ——— (1969). The validity of several plume rise formulas. *J. Air Poll. Control Assn.*, 19: 862–6.

Chan, W.H. and D.H.S. Chung (1986). Regional-scale precipitation scavenging of SO_2, SO_4, NO_3 and HNO_3. *Atmos. Environ.*, 20: 1397–402.

Clarke, R.H. (1970). Observational studies at the atmospheric boundary layer. *Q. J. R. Meteorol. Soc.*, 96: 91–114.

Corrsin, S. (1974). Limitations of gradient transport models in random walks and in turbulence. In *Advances in Geophysics*, 18A: 25–60.

Crutcher, H.L. (1984). Monitoring, sampling and managing meteorological data. In Randerson (1984), 136–46.

Csanady, G.T. (1955). Dispersal of dust particles from elevated sources. *Aust. J. Phys.*, 8: 545–50.

Dasch, J.M. (1986). Measurement of dry deposition to vegetation surfaces. *Water Air Soil Polln.*, 30: 205–10.

——— (1987). Measurement of dry deposition to surfaces in deciduous and pine canopies. *Envir. Pollut.*, 44: 261–77.

Davison, D.S., C.C. Fortems and K.L. Grandia (1977). The application of turbulence measurements to the dispersion of a large industrial effluent plume. *Joint Conference on Applications of Air Pollution Meteorology*, American Meteorological Society, 103–10.

Deacon, E.L. (1949). Vertical diffusion in the lowest layers of the atmosphere. *Q. J. R. Meteorol. Soc.*, 75: 89–103.

Deardorff, J.W. (1970). Convective velocity and temperature scales for the unstable planetary boundary layer and for Raleigh convection. *J. Atmos. Sci.*, 27: 1211–13.

Demerjian, K.L., J.A. Kerr and J.G. Calvert (1974). The mechanism of photochemical smog formation. *Adv. Env. Sci. Techn.*, 4: 1–262.

De Marrais, G. and N.F. Islitzer (1960). *Diffusion climatology of the national reactor testing station.* USAEC Report IDO-12015, Weather Bureau, Idaho Falls, Idaho.

Dockery, D.W. and J.D. Spengler (1981). Personal exposure to respirable particulates and sulfates. *J. Air Poll. Control Assn.*, 31: 153–9.

Donaldson, C. du P. (1973). Construction of a dynamic model of the production of atmospheric turbulence in the dispersal of atmospheric pollutants. In Haugen (1973), 313–92.

Drake, R.L., J.M. Hales, J. Mishima and D.R. Drewes (1979). Mathematical models for atmospheric pollutants. Appendix B: Chemical and physical properties of gases and aerosols. Electric Power Research Institute, EPRI EA-1131.

Draxler, R.R. (1976). Determination of atmospheric diffusion parameters. *Atmos. Environ.*, 10: 99–105.

——— (1987). Accuracy of various diffusion and stability schemes over Washington, D.C. *Atmos. Environ.*, 21: 491–9.

Drozdov, O.A. and A.A. Sepelevskij (1946). The theory of the interpolation of meteorological elements in a stochastic field and their application to questions of weather maps and network rationalization. *Trudy Niu Gugms*, Series No. 13.

Duewer, W.H., M.C. MacCracken and J.J. Walton (1978). The livermore regional air quality model: 2. Verification and sample application in the San Francisco Bay area. *J. Appl. Met.*, 11: 312–22.

Dyer, A.J. and B.B. Hicks (1970). Flux-gradient relationships in the constant flux layer. *Q. J. R. Meteorol. Soc.*, 96: 715–21.

Ehhalt, D., G. Pearman and I. Galbally, eds (1987). *Scientific Application of Baseline Observations of Atmospheric Composition (SABOAC).* Kluwer, 468 pp.

Ermak, D.L., S.T. Chan, D.L. Morgan and L.K. Morris (1982). A comparison of dense gas dispersion model simulations with Burro series LNG spill test results. *J. Hazardous Materials*, 6: 129–60.

Eschenroeder, A.Q. and J.R. Martinez (1972). Concepts and applications of photochemical smog models. In Gould (1972).

Eskinazi, D. and J.E. Cichanowicz (1989). Stationary combustion NO_x control, a

summary of the 1989 symposium. *J. Air Poll. Control Assn.*, 39: 1131–9.
Eskridge, R.E. and J.C.R. Hunt (1979). Highway modeling Part I: prediction of velocity and turbulence fields in the wake of vehicles. *J. Appl. Met.*, 18: 387–400.
——— and R.S. Thompson (1982). Experimental and theoretical study of the wake of a blockshaped vehicle in a shear-free boundary layer. *Atmos. Environ.*, 16: 2821–36.
Ferrari, L.M. and D.C. Johnson, eds (1984). *A Practical Guide to Sampling and Analysis*. Clean Air Society of Australia and New Zealand, NSW, Australia.
Ferris, B.G. (1978). Health effects of exposure to low levels of regulated pollutants – a critical review. *J. Air Poll. Control Assn.*, 28: 482–97.
Fielder, F. (1972). The effect of baroclinicity of the resistance law in a diabatic Ekman layer. *Beitr. Phys. Atm.*, 45: 164–73.
Fleischer, M.T. and F.L. Worley (1978). Orthogonal collocation – application to diffusion from point sources. *Atmos. Environ.*, 12: 1349–57.
Freney, J.R. and A.J. Nicolson (1980). *Sulfur in Australia*. Australian Academy of Science, Canberra City, 268 pp.
Friedlander, S.K. and J.H. Seinfeld (1969). A dynamic model of photochemical smog. *Env. Sci. Tech.*, 3: 1175–81.
Fritz, J.S. (1982). *Ion Chromatography*. Heidelberg Press, 203 pp.
Galloway, J. N., R.J. Charlson, M.O. Andreae and H. Rodhe, eds (1985). *The Biogeochemical Cycling of Sulfur and Nitrogen in the Remote Atmosphere*. Reidel, Dordrecht.
Gandin, L.S. (1965). *Objective analysis of meteorological fields*. Israeli Prog. for Scientific Translation, Jerusalem.
——— (1970). *The planning of meteorological station networks*. WMO Tech. Note No. 111, WMO, Geneva.
Geiger, R. (1965). *The climate near the ground*. Harvard University Press, Cambridge, 611 pp.
Georgopoulos, P.G. and J.H. Seinfeld (1988). Estimation of relative dispersion parameters from atmospheric turbulence spectra. *Atmos. Environ.*, 22: 31–41.
Gifford, F.A. (1961). Use of routine meteorological observations for estimating atmospheric dispersion. *Nuclear Safety*, 2: 47–51.
——— (1975). Atmospheric dispersion models for environmental pollution applications. In Haugen (1975), 35–58.
——— (1976). Turbulent diffusion typing schemes: a review. *Nuclear Safety*, 17: 68–86.
——— (1980). Smoke as a quantitative atmospheric diffusion tracer. *Atmos. Environ.*, 14: 1119–21.
——— (1982). Horizontal diffusion in the atmosphere: a Lagrangian-dynamical theory. *Atmos. Environ.*, 16: 505–12.
——— (1984). The random force theory: application to meso- and large-scale atmospheric diffusion. *Bound.-layer Meteorol.*, 30: 159–75.
——— (1987). The time scale of atmospheric diffusion considered in relation to the universal diffusion function, f_1. *Atmos. Environ.*, 21: 1315–20.
——— and S.R. Hanna (1973). Modelling urban air pollution. *Atmos. Environ.*, 7: 131–6.
Gilpin, A. (1971). *Air pollution*. University of Queensland Press, Brisbane, 67 pp.
Goddard, W.B. (1970). A floating drag-plate lysimeter, for atmospheric boundary

layer research. *J. Appl. Met.*, 9: 373–8.

Goldberg, E.D. (1982). *Atmospheric Chemistry: Report of the Dahlem Workshop on Atmospheric Chemistry*. Springer-Verlag, New York, 384 pp.

Golder, D. (1972). Relations among stability parameters in the surface layer. *Bound.-layer Meteorol.*, 3: 47–58.

Gould, R., ed. (1972). *Photochemical smog and ozone reactions*. ACS Adv. in Chemistry Series, American Chemical Society.

Graedel, T.E., D.T. Hawkins and L.D. Claxton (1986). *Atmospheric Chemical Compounds: Sources, Occurrence and Bioassay*. Academic Press.

Green, M.H. (1966). An air pollution index based on sulfur dioxide and smoke shade. *J. Air Poll. Control Assn.*, 11: 703–6.

Grefen, K. and J. Löbel, eds (1987). *Environmental Meteorology*. Kluwer, 670 pp.

Gregory, S. (1988). *Recent Climatic Change*. Belhaven Press, 326 pp.

Gryning, S.E., A.A.M. Holtslag, J.S. Irwin and B. Silversten (1987). Applied dispersion modelling based on meteorological scaling parameters. *Atmos. Environ.*, 21: 79–89.

——— P. van Ulden and S.E. Larsen (1983). Dispersion from a continuous ground level source investigated by a *K*-model. *Q. J. R. Meteorol. Soc.*, 109: 355–64.

Guderian, R. (1970a). Untersuchungen uber quantitative Beziehungen zwischen dem Schwelfelgehalt von Pflanzen und dem Schwefeldioxidegehalf der Luft – 1. *Z. Pflanzenkrankh. Pflanzenschutz*, 77: 200–20.

——— (1970b). Untersuchungen uber quantitative Beziehungen zwischen dem Schwelfelgehalt von Pflanzen und dem Schwefeldioxidegehalf der Luft – 2. *Z. Pflanzenkrankh. Pflanzenschutz*, 77: 289–308.

——— (1970c). Untersuchungen uber quantitative Beziehungen zwischen dem Schwelfelgehalt von Pflanzen und dem Schwefeldioxidegehalf der Luft – 3. *Z. Pflanzenkrankh. Pflanzenschutz*, 77: 387–99.

——— (1977). *Air pollution: phytotoxicity of acidic gases and its significance in air pollution control*. Springer Verlag, Berlin, 127 pp.

———, H. van Haut and H. Stratmann (1960). Probleme der Erfassung und Beurteilung von Wirkungen gasformiger Luftverunreinigungen auf die vegetation. *Z. Pflanzenkrankh. Pflanzenschutz*, 67: 257–64.

Hackney, J.D., W.S. Linn, R.D. Buckley, E.E. Pedersen, S.K. Karuze, D.C. Law and D.A. Fischer (1975a). Experimental studies on human health effects of air pollutants: 1 Design considerations. *Arch. Environ. Health*, 30: 373–8.

———, ———, J.G. Mohler, E.E. Pedersen, P. Breisacher and A. Russo (1975b). Experimental studies on human health effects of air pollutants: II Four-hour exposure to ozone alone and in combination with other pollutant gases. *Arch. Environ. Health*, 30: 379–84.

———, ———, D.C. Law, S.K. Karuza, H. Greenberg, R.D. Buckley and E.E. Pedersen (1975c). Experimental studies on human health effects of air pollutants: III Two-hour exposure to ozone alone and in combination with other pollutant gases. *Arch. Environ. Health*, 30: 385–90.

Hall, D.J. and S.L. Upton (1988). A wind tunnel study of the particle collection efficiency of an inverted Frisbee as a dust deposition gauge. *Atmos. Environ.*, 22: 1382–94.

Hanna, S.R. (1971). A simple method of calculating dispersion from urban area

sources. *J. Air Poll. Control Assn.*, 21: 774–7.
——— (1981). Lagrangian and Eulerian time-scale relations in the daytime boundary layer. *J. Appl. Met.*, 20: 242–9.
——— (1982). *Review of atmospheric diffusion models for regulatory applications.* WMO Tech. Note No. 177, WMO, Geneva, 37 pp.
——— (1988). Air quality model evaluation and uncertainty. *JAPCA*, 38: 406–12.
——— (1989). Confidence limits for air quality model evaluations, as estimated by bootstrap and jackknife resampling methods. *Atmos. Environ.*, 23: 1385–98.
———, G.A. Briggs, J. Deardorff, B.A. Egan, F.A. Gifford and F. Pasquill (1977). Summary of recommendations made by the AMS workshop on stability classification schemes and sigma curves. *Bull. Amer. Met. Soc.*, 58: 1305–9.
———, ——— and R.P. Hosker (1982). *Handbook on atmospheric diffusion.* Technical Information Centre, US Dept. of Energy, DE82002045 (DOE/TIC-11223), 102 pp.
———, T.V. Crawford, W.B. Bendel, J.W. Deardorff, T.W. Horst, G.H. Fichl, D. Randerson, S.P.S. Arya and J.M. Gorman (1978). Accuracy of dispersion models. *Bull. Amer. Met. Soc.*, 59: 1025–6.
——— and R.J. Paine (1987). Convective scaling applied to diffusion of buoyant plumes from tall stacks. *Atmos. Environ.*, 21: 2153–62.
Harrison, R.M. and R. Perry (1986). *Handbook of Air Pollution Analysis, 2nd edn.* Chapman and Hall and Methuen, New York, 634 pp.
Haugen, D.A., ed. (1973). *Workshop on micrometeorology.* American Meteorological Society, Boston, 392 pp.
———, ed. (1975). *Lectures on air pollution and environmental impact analyses.* American Meteorological Society, Boston, 296 pp.
Hazucha, M. and D.V. Bates (1975). Combined effect of ozone and sulfur dioxide on human pulmonary function. *Nature*, 257: 50–1.
Hecht, T.A. and J.H. Seinfeld (1972). Development and validation of a generalized mechanism for photochemical smog. *Env. Sci. Tech.*, 6: 47–57.
———, ——— and M.C. Dodge (1974). Further development of generalized mechanism for photochemical smog. *Env. Sci. Tech.*, 8: 327.
Hicks, B.B. (1976). Wind profile relationship from the Wangara experiment. *Q. J. R. Meteorol. Soc.*, 102: 535–51.
Hidy, G.M. (1984). *Aerosols, An Industrial and Environmental Science.* Academic Press, 774 pp.
Hileman, B. (1989). Global warming. *Chemical and Engineering News*, 67: 25–44.
Hoffnagle, G.F., M.E. Smith, T.V. Crawford and T.J. Lockhart (1981). On site meteorological instrumentation requirements to characterize diffusion from point sources – a workshop, 15–17 January 1980, Rayleigh, N.C. *Bull. Amer. Met. Soc.*, 62: 255–61.
Holland, J.Z. (1953). *A meteorological survey of the Oak Ridge area: final report covering the period 1948–1952.* USAEC Report ORO-99, Weather Bureau, Oak Ridge, Tenn.
Holtslag, A.A.M. and F.T.M. Nieuwstadt (1986). Scaling the atmospheric boundary layer. *Bound.-layer Meteorol.*, 36: 201–9.
——— and A.P. van Ulden (1983). A simple scheme for daytime estimates of the surface fluxes from routine weather data. *J. Climate Appl. Met.*, 22: 517–29.

Horst, T.W. (1980). A review of Gaussian diffusion-deposition models. In Shriner *et al.* (1980), 275–83.

Hosker, R.P. (1980). Practical application of air pollution deposition models - current status, data requirements and research needs. In Krupa and Legge (1980).

——— (1984). Flow and diffusion near obstacles. In Randerson (1984) 241–326.

Huang, C-Y. and S. Raman (1989). Application of the – E closure model to simulations of mesoscale topographic effects. *Bound.-layer Meteorol.*, 49: 169–95.

Huber, A.H. (1988). Performance of a Gaussian model for centreline concentrations in the wake of buildings. *Atmos. Environ.*, 22: 1039–50.

Husar, R.B. (1974). Recent developments in *in situ* size spectrum measurements of submicron aerosol. In *Instrumentation for monitoring air quality*, ASTRMSTP 555, American Society for Testing Materials, Philadelphia.

Idso, S.B. (1989). *Carbon dioxide and global change: earth in transition.* IBR Press, Institute for Biospheric Research, Inc., 631 E. Laguna Dr., Tempe, Arizona, 292 pp.

Inhaber, H. (1974). A set of suggested air quality indices for Canada. *Atmos. Environ.*, 9: 353–64.

Irwin, J.A. (1979). A theoretical variation of the wind profile power law exponent as a function of surface roughness and stability. *Atmos. Environ.*, 13: 191–4.

Irwin, J.S. (1983). Estimating plume dispersion - a comparison of several sigma schemes. *J. Climate Appl. Met.*, 22: 92–114.

Isaksen, I.S.A., ed. (1988). *Tropospheric ozone regional and global scale interactions.* Kluwer, 456 pp.

Ishizu, Y. (1980). General equation for the estimation of indoor pollution. *Environ. Sci. Technol.*, 14: 1254–7.

Jacobson, J.S. and A.C. Hill, eds (1970). *Recognition of air pollution injury to vegetation: a pictorial atlas.* Air Poll. Cont. Assn., Pittsburg, 1122 pp.

Jakeman, A.J. and R.W. Simpson (1983). The application of a risk assessment strategy to determine air pollution modeling and monitoring policy. *Science and Public Policy*, 6: 289–94.

Johnson, W.B. (1983). Meteorological tracer techniques for parameterizing atmospheric dispersion. *J. Climate Appl. Met.*, 22: 931–46.

——— and R.E. Ruff (1975). Observational systems and techniques in air pollution meteorology. In Haugen (1975), 243–74.

Johnson, R.W. and G.E. Gordon, eds (1987). *The chemistry of acid rain, sources and atmospheric processes: ACS Symposium Series 349.* American Chemical Society, Washington, D.C., 337 pp.

Junge, C.E. (1963). *Air chemistry and radioactivity.* Academic Press, New York, pp. 291–8.

——— (1972). The cycle of atmospheric trace gases — natural and manmade. *Q. J. R. Meteorol. Soc.*, 98: 711–29.

——— (1974). Residence time and variability of tropospheric trace gases. *Tellus*, 26: 477–88.

Kagawa, J. and T. Toyama (1975). Photochemical air pollution: its effect on respiratory function of elementary school children. *Arch. Environ. Health*, 30: 117–22.

KAMS (1982). *The Kwinama Air Modelling Study.* Report 10, Department of Conservation and the Environment, Perth, Western Australia, 96 pp.

Kamst, F.H. and T.J. Lyons (1982a). The evaluation of diffusivities for a non-uniform site. *Atmos. Environ.*, 16: 379–90.

——— and ——— (1982b). A regional air quality model for the Kwinana industrial area of Western Australia. *Atmos. Environ.*, 16: 401–12.

Katz, M., ed. (1977). *Methods of air sampling and analysis.* American Public Health Assoc. Washington, DC, 984 pp.

——— (1980). Advances in the analysis of air contaminants a critical review. *J. Air Poll. Control Assn.*, 30: 528.

Keith, L.H. (1987). *Principles of Environmental Sampling.* American Chemical Society, Washington, DC, 458 pp.

Kerman, B.R., R.E. Nickle, R.V. Portelli and N.B. Trivett (1982). The Nanticoke shoreline diffusion experiment, June 1978 – II – Internal boundary layer structure, *Atmos. Environ.*, 16: 423–37.

Kleindorfer, P.R. and H.C. Kunreuther, eds (1987). *Insuring and Managing Hazardous Risks: From Seveso to Bhopal and Beyond.* Springer, Berlin.

Kondo, J., O. Kanechika and N. Yasuda (1978). Heat and momentum transfers under strong stability in the atmospheric surface layer. *J. Atmos. Sci.*, 35: 1012–21.

——— and H. Yamazawa (1986). Aerodynamic roughness over an inhomogeneous ground surface. *Bound.-layer Meteorol.*, 35: 331–48.

Kondratyev, K. Ya. (1969). *Radiation in the atmosphere.* Academic Press, New York, 912 pp.

Koopman, R.P., D.L. Ermak and S.T. Chan (1989). A review of recent field tests and mathematical modelling of atmospheric dispersion of large spills of denser than air gases. *Atmos. Environ.*, 23: 731–45.

Krupa, S.V. and A.H. Legge, eds (1980). *Proc. of International Conference on Air Pollutants and their effects on the terrestrial ecosystem.* Wiley and Sons, New York.

Langstaff, J., C. Seigneur and M.-K. Liu (1987). Design of an optimum air monitoring network for exposure assessments. *Atmos. Environ.*, 21: 1393–410.

Lefohn, A.S. and S.V. Krupa (1988). Acid precipitation, a technical amplification of NAPAP's findings. *J. Air Poll. Control Assn.*, 38: 766–76.

Legros, Ch. and A.L. Berger (1978). Sensitivity analysis of a Gaussian plume model. in *WMO Symposium on boundary layer physics applied to specific problems in air pollution.* Norrkoping, June 1978, 169–74.

Lenschow, D.H., ed. (1986). *Probing the atmospheric boundary layer.* American Meteorological Society, Boston, 269 pp.

Lettau, H. (1950). A re-examination of the 'Leipzig wind profile' considering some relations between wind and turbulence in the friction layer. *Tellus*, 2: 125–9.

——— (1970). Physical and meteorological basis for mathematical models of urban diffusion. *Proc. Symposium on Multiple Source Urban Diffusion Models*, USEPA, Air Pollution Control Publication No. AP 86.

Lim, C.K. (1986). *HPLC of Small Molecules: A Practical Approach.* IRL Press, Oxford, 333 pp.

Linn, W.S., J.D. Hackney, E.E. Pedersen, P. Breisacher, J.V. Patterson, C.A. Mulry and J.F. Coyle (1976). Respiratory function and symptoms in urban office workers in relation to oxidant air pollution exposure. *Am. Rev. Resp. Disease*, 114: 477–83.

Llewellen, W.S. and M.E. Teske (1976a). Second-order closure modelling of diffusion in the atmospheric boundary layer. *Bound.-layer Meteorol.*, 10: 69–90.

——— and ——— (1976b). A second-order closure model of turbulent transport in the coastal planetary boundary layer. *Conference on Coastal Meteorology*, American Meteorological Society, 21–3.

Lumley, J.L. and H.A. Panofsky (1964). *The structure of atmospheric turbulence.* Wiley Interscience, New York, 239 pp.

Luna, R.E. and H.W. Church (1971). A comparison of turbulence intensity and stability ratio measurements to Pasquill turbulence types. *Conference on Air Pollution Meteorology*, American Meteorological Society.

Lyons, T.J., F.H. Kamst and I.D. Watson (1982). A simple sonde system for use in air pollution studies. *Atmos. Environ.*, 16: 391–9.

——— R.O. Pitts, J.A. Blockley, J.R. Kenworthy and P.W.G. Newman (1990). Motor vehicle emission inventory for the Perth airshed. *J. Roy. Soc. West Aust.*, 72: 67–74.

——— and R.K. Steedman (1981). Stagnation and nocturnal temperature jumps in a desert region of low relief. *Bound.-layer Meteorol.*, 21: 369–87.

Lyons, W.A. (1975). Turbulent diffusion and pollutant transport in shoreline environments. In Haugen (1975), 136–208.

Maas, S.J. and P.P. Harrison (1977). Dispersion over water: a case study of a non-buoyant plume in the Santa Barbara Channel, California. *Joint Conference on Applications of Air Pollution Meteorology*, American Meteorological Society, 12–15.

McCartney, E.J. (1976). *Optics of the Atmosphere, Scattering by Molecules and Particles.* Wiley, 408 pp.

MacCracken, M.C., D.J. Wuebbles, J.J. Walton, W.H. Duewer and K.E. Grant (1978). The Livermore regional air quality model: 1. Concept and development. *J. Appl. Met.*, 17: 254–72.

McCormac, B.M., ed. (1971). *Introduction to the scientific study of atmospheric pollution.* Reidel, Dordrecht, 169 pp.

McLaughlin, S.B. (1985). Effect of air pollution on forests. *J. Air Poll. Control Assn.*, 35: 512–34.

McNider, R.T. (1983). Energy conservation in Lagrangian-conditioned particles models. *Sixth symposium on turbulence and diffusion*, American Meteorological Society, Boston, 51–4.

McVehil, G.E. (1964). Wind and temperature profiles near the ground in stable stratification. *Q. J. R. Meteorol. Soc.*, 90: 136–46.

Mahrt, L. (1987). Grid-averaged surface fluxes. *Mon. Weather Rev.*, 115: 1550–60.

Mainwaring, S. (1989). Air quality — a view from the sidelines. *Clean Air*, 23: 24–9.

Makino, K. and I. Mizoguchi (1975). Symptoms caused by photochemical smog. *Japn. J. Public Health*, 22: 421–30.

Mason, C.J. and H. Moses (1984). Meteorological instrumentation. In Randerson (1984), 81–135.

Mathy, P., ed. (1987). *Air Pollution and Ecosystems.* Kluwer, 1004 pp.

Maul, P.R. (1978). The effect of the turning of the wind vector with height on the ground level trajectory of a diffusing cloud. *Atmos. Environ.*, 12: 1045–50.

Miller, C.W. (1978). An examination of Gaussian plume dispersion parameters for rough terrain. *Atmos. Environ.*, 12: 1359–64.

Misra, P.K., W.H. Chan, D. Chung and A.J.S. Tang (1985). Scavenging ratios of acidic pollutants and their use in long range transport models. *Atmos. Environ.*, 19: 1471–5.

Moeng, C-H. and J.C. Wyngaard (1986). An analysis of closures for pressure-scalar covariances in the convective boundary layer. *J. Atmos. Sci.*, 43: 2499–513.

——— and ——— (1989). Evaluation of turbulent transport and dissipation closures in second-order modelling. *J. Atmos. Sci.*, 46: 2311–30.

Monin, A.S. and A.M. Obukhov (1954). The basic laws of turbulent mixing in the surface layer of the atmosphere. *Akad. Nauk. SSSR. Trud. Geofiz. Inst.*, 24 (151): 163–87.

Munn, R.E. (1966). *Descriptive micrometeorology.* Academic Press, New York, 245 pp.

——— (1970). *Biometeorological methods.* Academic Press, New York, 336 pp.

——— (1973). Urban meteorology: some selected topics. *Bull. Amer. Met. Soc.*, 54: 90.

———, ed. (1975). *Environmental impact assessment: principles and procedures.* SCOPE Report 5, United Nations Environmental Program, 173 pp.

Murray, F. (1989). Acid rain and acid gases in Australia. *Arch. Environmental Contamination and Toxicology*, 18, 131–6.

——— and S. Wilson (1989). The relationship between sulfur dioxide concentration and yield of five crops in Australia. *Clean Air*, 23: 24–9.

Nakamori, Y., S. Ikeda and Y. Sawaragi (1979). Design of air pollution monitoring system by spatial sample stratification. *Atmos. Environ.*, 13: 97–103.

Nero, A.V. Jr. (1988). Estimated risk of lung cancer from exposure to radon decay products in US homes: a brief review. *Atmos. Environ.*, 22: 2205–11.

Nicholson, K.W. (1988a). A review of particle resuspension. *Atmos. Environ.*, 22: 2639–51.

——— (1988b). The dry deposition of small particles: a review of experimental results. *Atmos. Environ.*, 22: 2653–66.

Nieuwstadt, F.T.M. (1984). The turbulent structure of the stable nocturnal boundary layer. *J. Atmos. Sci.*, 41: 2202–16.

——— and H. van Dop (1982). *Atmospheric turbulence and air pollution modelling.* Reidel, Dordrecht, 358 pp.

Niewiadomski, M. (1989). Sulphur dioxide and sulphate in a three-dimensional field of convective clouds: numerical simulations. *Atmos. Environ.*, 23: 477–87.

Noll, K.E. and T.L. Miller (1977). *Air monitoring survey design.* Ann Arbor Science, Ann Arbor, 296 pp.

Nordo, J. (1976). Long range transport of air pollutants in Europe and acid precipitation in Norway. *Water, Air and Soil Pollution*, 6: 199–217.

O'Gara, P.J. (1922). Sulphur dioxide and fume problems and their solutions. *Ind. Eng. Chem.*, 14: 744 pp.

Oke, T.R. (1978). *Boundary layer climates.* Methuen, London, 372 pp.

O'Riordan, T. and B. Wynne (1987). Regulating environmental risks: a comparative perspective. In Kleindorfer and Kunreuther (1987), 389–410.

Ott, W.R. (1978). *Environmental indices: theory and practice.* Ann Arbor Science, Ann Arbor, 371 pp.

——— and G.C. Thom (1976). A critical review of air pollution index systems in the United States and Canada. *J. Air Poll. Control Assn.*, 26: 460–70.

Overcamp, T.J. (1976). A general Gaussian diffusion–deposition model for elevated point sources. *J. Appl. Met.*, 15: 1167–71.

Padmanbhamurty, B. and V.P. Subrahmanyam (1961). Diurnal and seasonal variations of heat-flow into soil at Waltair. *Indian J. Meteorol. Geophys.*, 12: 261–6.

Pasquill, F.(1961). The estimation of the dispersion of windborne material. *Met. Mag.*, 90: 33–49.

——— (1974). *Atmospheric diffusion*. Ellis Horwood Ltd., Chichester, 429 pp.

——— (1975). The dispersion of material in the atmospheric boundary layer: the basis for generalization. In Haugen (1975), 1–34.

——— (1976a). *The 'Gaussian-plume' model with limited vertical mixing*. US Environmental Protection Agency, Research Triangle Park, NC. EPO – 600/4-76-042.

——— (1976b). *Atmospheric dispersion parameters in Gaussian plume modelling – Part II: possible requirements for change in the Turner workbook values*. US Environmental Protection Agency, Research Triangle Park, NC. EPA – 600/4-76-0306.

——— and F.B. Smith (1983). *Atmospheric diffusion* (3rd edn). Wiley, New York, 437 pp.

Pearman, G.F. (1989). *Greenhouse, Planning for Climatic Change*. CSIRO Publications, East Melbourne, Victoria, Australia, 760 pp.

Petersen, W.B. (1978). *User's guide for PAL – a Gaussian-plume algorithm for point, area and line sources*. US Environmental Protection Agency, Research Triangle Park, NC. EPA – 600/4-78-012 (NTIS Accession Number PB-281 306).

Pielke, R.A. (1984). *Mesoscale meteorological modeling*. Academic Press, New York, 612 pp.

———, R.T. McNider, M. Segal and Y. Mahrer (1983). The use of a mesoscale numerical model for evaluations of pollutant transport and diffusion in coastal regions and over irregular terrain. *Bull. Amer. Met. Soc.*, 64: 243–9.

Pierce, T.E. and D.B. Turner (1980). *User's guide for MPTER: a multiple point Gaussian dispersion algorithm with optional terrain adjustment*. US Environmental Protection Agency, Research Triangle Park, NC. EPA – 600/8-80-016 (NTIS Accession Number PB80-197 361).

——— and ———, J.A. Catalano and F.V. Hale (1982). *PTPLU – a single source Gaussian dispersion algorithm – user's guide*. US Environmental Protection Agency, Research Triangle Park, NC. EPA – 600/8-82-014 (NTIS Accession Number PB83-211 235).

Plate, E.J., ed. (1982). *Engineering meteorology*. Elsevier, Amsterdam, 740 pp.

Pollak, L.W. and A.L. Metnieks (1960). Intrinsic calibration of the photo-electric nucleus counter Model 1957 with convergent light-beam. Air Force Cambridge Research Laboratory Technical (Scientific) Note No. 9, Contract AF 61 (052)-26.

Ragland, K.W. and R.L. Dennis (1975). Point source atmospheric diffusion model with variable wind and diffusivity profiles. *Atmos. Environ.*, 9: 175–89.

Randerson, D., ed. (1984). *Atmospheric Science and Power Production*. US Department of Energy, DOE/TIC-27601 (DE 84005177), 850 pp.

Rao, K.S. (1982). *Analytical solutions of a gradient-transfer model for plume deposition and sedimentation*. US Environmental Protection Agency, Research

Triangle Park, NC. EPA – 600/8-82-079 (NTIS Accession Number PB82-215 153).

——— and L. Satterfield(1982). *MPTER-DS: the MPTER model including deposition and sedimentation*. US Environmental Protection Agency, Research Triangle Park, NC. EPA – 600/8-82-024 (NTIS Accession Number PB83-114 207).

——— and H.F. Snodgrass (1982). *PAL-DS Model: the PAL model including deposition and sedimentation – user's guide*. US Environmental Protection Agency, Research Triangle Park, NC. EPA – 600/8-82-023 (NTIS Accession Number PB83-117 739).

Rayner, K.N. (1987). Dispersion of atmospheric pollutants from point sources in a coastal environment. Doctoral dissertation, Environmental Sciences, Murdoch University, 249 pp.

Reid, J.D. (1979). Markov chain simulations of vertical dispersion in neutral surface layer for surface and elevated releases. *Bound.-layer Meteorol.*, 16: 3–22.

Renn, O. (1989). Risk communication at the community level: European lessons from the Seveso directive. *JAPCA*, 39: 1301–8.

Rich, T.A. (1967). Air pollution studies aided by overall air pollution index. *Environ. Sci. and Technol.*, 1: 796–800.

Richard, L.W. (1988). Sight path measurements for visibility monitoring and research. *J. Air Poll. Control Assn.*, 38: 784–91.

Rider, N.E. (1954). Eddy diffusion of momentum, water vapour and heat near the ground. *Phil. Transact. Roy. Soc.*, A246: 481–501.

Roberts, D.B. and D.J. Williams (1978). Atmospheric oxidation of sulfur dioxide. In *Proceedings of the 6th International Clean Air Conference*, ed. E.T. White. Ann Arbor Science Publishers, 780 pp.

Rodhe, H. and R. Herrera, eds (1987). *Acidification in tropical Countries*. Wiley, 68 pp.

Rosenberg, N.J. (1974). *Microclimate: the biological environment* Wiley, New York, 315 pp.

Ruskin, R.E. and W.D. Scott (1974). Weather modification instruments and their use. In *Weather and Climate Modification*, ed. W.N. Hess. J. Wiley, New York, p. 181.

Ryan, P.B. J.D. Spengler and P.F. Halfpenny (1988). Sequential box models for indoor air quality: application to airliner cabin air quality. *Atmos. Environ.*, 22: 1031–8.

———, ——— and R. Letz (1983). The effects of kerosene heaters on indoor pollutant concentrations: a monitoring and modeling study. *Atmos. Environ.*, 17: 1339–45.

Savoie, D.L., J.M. Prospero and R.T. Nees (1987). Washout ratios of nitrate, non-sea-salt sulfate and sea-salt on Virginia Key, Florida and on American Samoa. *Atmos. Environ.*, 21: 103–12.

Sawford, B.L. (1984). The basis for, and some limitations of, the Langevin equation in atmospheric relative dispersion modelling. *Atmos. Environ.*, 18: 2405–11.

——— and D.G. Ross (1985). Workshop on regulatory air quality modelling in Australia — 8th International Clean Air Conference. *Clean Air*, 19: 82–7.

Schoettlin, C.E. and E. Landan (1961). Air pollution and asthmatic attacks in the Los Angeles area. *Public Health Reports*, 76: 545–8.

Schulman, L.L. and J.S. Scire (1980a). *Development of an air quality dispersion* model for aluminum reduction plants. Environmental Research and Technology, Inc., Concord, Ma., Document P-7304A (NTIS Accession Number PB81-164 634).

——— and ——— (1980b). *Buoyant line and point source (BLP) dispersion model user's guide*. Environmental Research and Technology, Inc., Concord, Ma., Document P-7304B (NTIS Accession Number PB81-164 642).

Scott, B.C. (1982). Predictions of in-cloud conversion rates of SO_2 to SO_4 based upon a simple chemical and kinematic storm model. *Atmos. Environ.*, 16: 1735–52.

Scott, W.D., and P.V. Hobbs (1967). The formation of sulfate in water droplets. *J. Atmos. Sci.*, 24: 54–7.

Seinfeld, J.H. (1972). Optimal location of pollutant monitoring stations in an airshed. *Atmos. Environ.*, 6: 847–58.

——— (1986). *Atmospheric Chemistry and Physics of Air Pollution*. Wiley Interscience, New York, 738 pp.

Sellers, W.D. (1965). *Physical climatology*. University of Chicago Press, Chicago, 272 pp.

Selway, M.D., R.J. Allen and R.A. Wadden (1980). Ozone production from photocopying machines. *Am. Ind. Hyg. Assn. J.*, 41: 455–9.

Sexton K. and J.J. Wesolowski (1985). Safeguarding indoor air quality. *Environ. Sci. Technol.*, 19: 305–9.

Shair, F.H. and K.L. Heitner (1974). Theoretical model for relating indoor pollutant concentrations to those outside. *Env. Sci. Tech.*, 8: 444–51.

Shanley, J.B. (1989). Field measurements of dry deposition to spruce foliage and petri dishes in the Black Forest, FRG. *Atmos. Environ.*, 23: 403–14.

Sheih, C.M., G.D. Hess and B.B. Hicks (1978). Design of network experiments for regional-scale atmospheric pollutant transport and transformation. *Atmos. Environ.*, 12: 1745–53.

Shenfeld, L. (1970). Note on Ontario's air pollution index and alert system. *J. Air Poll. Control Assn.*, 20: 622.

Sheppard, P.A., D.T. Tribble and J.R. Garratt (1972). Studies of turbulence in the surface layer over water (Lough Neagh) Part I. Instrumentation programme profiles. *Q. J. R. Meteorol. Soc.*, 98: 627–41.

Shir, C.C. (1973). A preliminary numerical study of atmospheric turbulent flows in the idealized planetary boundary layer. *J. Atmos. Sci.*, 30: 1327–39.

——— and L.J. Shieh (1974). A generalized urban air pollution model and its application to the study of SO_2 distributions in the St Louis Metropolitan area. *J. Appl. Met.*, 13: 185–204.

Shriner, D.S., C.R. Richmond and S.E. Lindberg, eds (1980). *Atmospheric sulfur deposition, environmental impact and health effects*. Ann Arbor Science, Ann Arbor, Michigan.

Singer, J.A. and G.S. Raynor (1957). *Analysis of meteorological tower data, April 1950–March 1952*. USAEC Report BNL-461, Brookhaven National Laboratory.

Skoog, D.A. (1985). *Principles of Instrumental Analysis*. Saunders, Japan, 948 pp.

Slade, D.H., ed. (1968). *Meteorology and atomic energy – 1968*. USAEC, TID-24190, 445 pp.

Smith, F.B. and G.H. Jeffrey (1973). The prediction of high concentrations of SO_2

in London and Manchester air. *Proc. 3rd Meeting Expert Panel on Air Pollution Modelling*. NATO/CCMS, Paris, 322 pp.

Smith, M.E. (1984). Review of the attributes and performance of 10 rural diffusion models. *Bull. Amer. Met. Soc.*, 65: 554–8.

Sneyers, R. (1973). Sur la densité optimale des réseaux météorologiques. *Arch. Met. Geoph. Biokl.*, Ser. B, 21: 17.

Stalker, W.W., R.C. Dickerson and G.D. Kramer (1962). Sampling station and time requirements for urban air pollution surveys, Part IV. *J. Air Poll. Control Assn.*, 12: 361–75.

Start, G.E. and L.L. Wendell (1974). *Regional effluent dispersion calculations considering spatial and temporal meteorological variations*. NOAA Tech. Memo. ERL ARL-44, Air Resources Lab., Idaho Falls, 63 pp.

Steffen, W.L. and O.T. Denmead, eds. (1988). *Flow and Transport in the Natural Environment: Advances and Applications*. Springer-Verlag, Berlin, 384 pp.

Stern, A.C., R.W. Boubel, D.B. Turner and D.L. Fox (1984). *Fundamentals of Air Pollution*, Academic Press, London, 530 pp.

———, H.C. Wohlers, R.W. Boubel and W.P. Lowry (1973). *Fundamentals of Air Pollution*, Academic Press, New York, 492 pp.

Stull, R.B. (1988). *An introduction to boundary layer meteorology*. Kluwer Academic, Dordrecht, 666 pp.

Sutton, O.G. (1953). *Micrometeorology*. McGraw-Hill, New York, 332 pp.

Swinbank, W.C. (1968). A comparison between predictions of dimensional analysis for the constant flux layer and observations in unstable conditions. *Q. J. R. Meteorol. Soc.*, 94: 460–7.

——— and A.J. Dyer (1967). An experimental study in micrometeorology. *Q. J. R. Meteorol. Soc.*, 93: 494–500.

Tangermann, G. (1978). Numerical simulations of air pollutant dispersion in a stratified planetary boundary layer. *Atmos. Environ.*, 12: 1365–9.

Taylor, G.I. (1938). The spectrum of turbulence. *Proc. Roy. Soc.*, Ser. A. 164: 453–65.

Taylor, J.K. (1987). *Quality Assurance of Chemical Measurements*. Lewis Publishers, Inc., Chelsea, MI., 328 pp.

ten Brink, H.M., A.J. Janssen and J. Slanina (1988). Plume wash-out near a coal-fired power plant: measurements and model calculations. *Atmos. Environ.*, 22: 177–87.

Thom, G.C. and W.R. Ott (1976). A proposed uniform air pollution index. *Atmos. Environ.*, 10: 261–4.

Thomas, M.D. and G.R. Hill (1935). Absorption of sulphur dioxide by alfalfa and its relation to leaf injury. *Plant Physiol.*, 10: 291–307.

———, R.H. Hendricks and G.R. Hill (1944a). Some chemical reactions of sulfur dioxide after absorption by alfalfa and sugar beet. *Plant Physiol.*, 19: 212–26.

———, ———, C.C. Bryner and G.R. Hill (1944b). A study of the sulphur metabolism of wheat barley and corn using radioactive sulfur. *Plant Physiol.*, 19: 227–44.

Thomson, D.J. (1984). Random walk modelling of diffusion in inhomogenous turbulence. *Q. J. R. Meteorol. Soc.*, 110: 1107–20.

——— (1985). A random walk model for dispersion in turbulent flows and its application to dispersion in a valley. *Q. J. R. Meteorol. Soc.*, 11: 511–30.

Toussaint, L.F. (1980). Optimal location of sensors using the Kalman filter. M.Sc. thesis, Murdoch University, 93 pp.

Turner, D.B. (1967). *Workbook of atmospheric dispersion estimates.* Public Health Service, Pub. 999-AP-26, R.A. Taft Sanitary Engineering Centre, Cinncincinnati, Ohio.

——— (1979). Atmospheric dispersion modelling: a critical review. *J. Air Poll. Control Assn.*, 29: 502–19.

——— and A.D. Busse (1973). *User's guide to the interactive versions of three point source dispersion programs: PTMAX, PTDIS and PTMTP.* US Environmental Protection Agency, Research Triangle Park, NC. (NTIS Accession Number PB81-164 667).

——— and J.H. Novak (1978a). *User's guide for RAM Volume 1. Algorithm description and use.* US Environmental Protection Agency, Research Triangle Park, NC. EPA – 600/8-78-016A (NTIS Accession Number PB-294 791).

——— and ——— (1978b). *User's guide for RAM Volume 2. Data preparation and listings.* US Environmental Protection Agency, Research Triangle Park, NC. EPA – 600/8-78-016B (NTIS Accession Number PB-294 792).

Twomey, S. (1977). *Atmospheric aerosols.* Elsevier Scientific, Amsterdam.

Van der Hoven, I. (1968). Deposition of particles and gases. In Slade (1968), 202–7.

Van Haut, H. (1961). Die Analyse von Schwefeldioxidwirkungen auf Pflanzen im Laboratoriumsversuch. *Staub*, 21: 52–6.

Venkatram, A. (1988a). An interpretation of Taylor's statistical analysis of particle dispersion. *Atmos. Environ.*, 22: 865–8.

——— (1988b). Inherent uncertainty in air quality modelling. *Atmos. Environ.*, 22: 1221–7.

——— (1988c). Dispersion in the stable boundary layer. In Venkatram and Wyngaard (1988), 229–65.

——— (1988d). Topics in applied dispersion modelling. In Venkatram and Wyngaard (1988), 267–324.

———, P.K. Karamchandani and P.K. Misra (1988). Testing a comprehensive acid deposition model. *Atmos. Environ.*, 22: 737–47.

——— and R. Paine. (1985). A model to estimate dispersion of elevated releases into a shear dominated boundary layer. *Atmos. Environ.*, 19: 1797–1805.

——— and J.C. Wyngaard, eds (1988). *Lectures on air pollution modeling.* American Meteorological Society, Boston, 390 pp.

Wadden, R.A. and P.A. Scheff (1983). *Indoor air pollution.* Wiley Interscience, New York, 213 pp.

Walker, E.R. (1963). *Atmospheric turbulence characteristics measured at Suffield Experimental Station.* Defence Research Board Project No. D52-32-01-04, Suffield Experimental Station, Dept. of National Defence of Canada.

Wallace, J.M. and P.V. Hobbs (1977). *Atmospheric science – an introductory survey.* Academic Press, New York, 467 pp.

Wayne, R.P. (1987). The photochemistry of ozone. *Atmos. Environ.*, 21: 1683–94.

Webb, E.K. (1970). Profile relationships: the log-linear range and extension to strong stability. *Q. J. R. Meteorol. Soc.*, 96: 67–90.

Weber, A.H. (1976). *Atmospheric dispersion parameters in Gaussian plume modelling. Part I: Review of current systems and possible future developments.* EPA – 600/4-76-030a, 58 pp.

Weil, J.C. (1988a). Plume rise. In Venkatram and Wyngaard (1988), 119–66.

——— (1988b). Dispersion in the convective boundary layer. In Venkatram and Wyngaard (1988), 167–227.

——— (1988c). Atmospheric dispersion – observations and models. In Steffen and Denmead (1988), 352–76.

——— and R.P. Brower (1984). An updated Gaussian plume model for tall stacks. *J. Air Poll. Control Assn.*, 34: 818–27.

Weisz, H. (1970). *Microanalysis by the ring-oven technique.* Pergamon Press, Oxford, 170 pp.

Wengle, H., B. van den Bosch and J.H. Seinfeld (1978). Solution of atmospheric diffusion problems by pseudo-spectral and orthogonal collocation methods. *Atmos. Environ.*, 12: 1021–32.

Wesely, M.L. (1989). Parameterization of surface resistances to gaseous dry deposition in regional-scale numerical models. *Atmos. Environ.*, 23: 1293–304.

Wetzel, P.J. and J-T. Chang (1988). Evapotranspiration from non-uniform surfaces. A first approach for short-term numerical weather prediction. *Mon. Weather Rev.*, 116: 600–21.

Whitby, K.T. (1975). *Modelling of atmospheric aerosol particle size distributions.* Particle Technology Lab. Publ. 253, University of Minnesota, Minneapolis.

Wieringa, J. (1980). Representativeness of wind observations at airports. *Bull. Amer. Met. Soc.*, 61: 962–71.

Woodward, J.L., J.A. Havens, W.C. McBride and J.R. Taft (1982). A comparison with experimental data of several models for dispersion of heavy vapour clouds. *J. Hazardous Materials*, 6: 161–80.

Wuebbles, D.J., K.E. Grant, P.S. Connell and J.E. Penner (1989). The role of atmospheric chemistry in climate change. *J. Air Poll. Control Assn.*, 39: 22–8.

Wyngaard, J.C. (1985). Structure of the planetary boundary layer and implications for its modelling. *J. Climate Appl. Met.*, 24: 1131–42.

Wyngaard. J.C. (1988). Structure of the planetary boundary layer. In Venkatram and Wyngaard (1988), 9–61.

Yamada, T. and G.L. Mellor (1975). A simulation of the Wangara atmospheric boundary layer data. *J. Atmos. Sci.*, 32: 2309–29.

Yassky, D. (1983). Estimation of dispersion parameters from photographic density measurements on smoke puffs. *Atmos. Environ.*, 17: 283–90.

Young, J.W.S. (1982). From emission to deposition: modelling and source–receptor relationships. *Proceedings Atmospheric Deposition Speciality Conference*, Air Pollution Control Association, SP-49, 46–81.

Zeller, K.F. (1984). The environmental impact statement. In Randerson (1984), 789–809.

Zib, P. (1977). *Urban air pollution dispersion models: a critical survey.* Dept. of Geography and Environmental Studies, University of Witwatersrand, Johannesburg, Occasional Paper No. 16, 44 pp.

Ziegler, J. (1975). The effect of SO_2 pollution on plant metabolism. *Residue Reviews*, 56: 79–105.

Zimmerman, J.R. and R.S. Thompson (1975). *User's guide for HIWAY: a highway air pollution model.* US Environmental Protection Agency, Research Triangle Park, NC. EPS – 650/4-74-008 (NTIS Accession Number PB-239 444).

INDEX